Science and Relativism

Science and Its Conceptual Foundations

David L. Hull, Editor

Science
AND
Relativism

Some Key Controversies
in the
Philosophy of Science

Larry Laudan

THE UNIVERSITY OF CHICAGO PRESS

Chicago and London

The University of Chicago Press, Chicago 60637
The University of Chicago Press, Ltd., London

Printed in the United States of America

99 98 97 96 95 94 93 92 5 4 3

Library of Congress Cataloging-in-Publication Data

Laudan, Larry.
Science and relativism : some key controversies in the philosophy of science/Larry Laudan.
p. cm. — (Science and its conceptual foundations)
Includes bibliographical references.
ISBN 0-226-46948-4 (cloth) : — ISBN 0-226-46949-2 (paper)
1. Science—Philosophy. 2. Science—History. 3. Relativity.
I. Title. II. Series.
Q175.L294 1990
149′.73—dc20 90-32112
CIP

The paper used in this publication meets the minimum requirements of the American National Standard for Information Sciences—Permanence of Paper for Printed Library Materials, ANSI Z39.48–1984.

Contents

Preface
vii

Note to the Reader
xiii

1. Progress and Cumulativity
1

2. Theory-ladenness and Underdetermination
33

3. Holism
69

4. Standards of Success
93

5. Incommensurability
121

6. Interests and the Social Determinants of Belief
146

References
171

Index
175

Preface

Philosophers of science, my friends keep telling me, do not write dialogues. We fancy ourselves a hard-bitten lot, who reckon that dialogues are either too effete or too diffuse to lend themselves to the sort of rigor for which, in our vainer moments, we pride ourselves. That reaction is doubtless too hasty, for some of the brightest gems of arguments in the history of science and philosophy can be found in the dialogues of Plato, Galileo, Berkeley, and Hume. I do not pretend to be in that league but I do find myself confronted by the same sort of expository problem those thinkers faced. Galileo, for instance, bemoaned the fact that those outside his own specialty of mechanics found the technical treatises of his discipline inaccessible. Still worse, there was widespread confusion among his (nonscientific) contemporaries about what the "new" science did and did not amount to. He decided that a dialogue in the vernacular might put the situation right. Similarly, it is plausible to suppose that Plato's surviving dialogues were an effort at rendering more intelligible to those not in the Academy the arcane treatises being debated among its members.

I confront a similar perplexity. Those outside philosophy of science proper—and I include here many scientists (both natural and social), as well as philosophers in other specialties—commonly have a certain picture of the recent history of my subject and of its current status. According to that picture, positivism reigned supreme for a century, from Comte to Carnap. Then, so the story goes, in the early 1960s, positivism was overthrown and replaced by what has uniformly come to be called (by everyone *except* specialists in the field) "post-positivist philosophy of science." Not only is there broad consensus outside the philosophy of science that this revolution occurred; there is also broad agreement that this revolution rendered problematic—perhaps beyond hope of redemption—such key notions as progress, objectivity, and rationality. In sum, many who are not philosophers of science (from cultural philosophers like Rorty and Winch to sociologists like Barnes and Collins) appear to believe that contemporary philosophy of science provides potent arguments on behalf of a *radical relativism* about knowledge in general and scientific knowledge in particular. Relativism has many nuances, some of which will be explored below. But it can be defined, to a first order of approximation, as the thesis that the natural world and such evidence as we have about that world do little or nothing to constrain our beliefs. In a phrase, the relativists' slogan is "The way we take things to be is quite independent of the way things are." It is this view that many current writers take away from the study of philosophy of science.

My belief, by contrast, is that strong forms of epistemic relativism derive scant support from a clearheaded understanding of the contemporary state of the art in philosophy of science. I am not alone in that conviction; most of my fellow philosophers of science would doubtless wholeheartedly concur. But that consensus within the discipline apparently cuts little ice with those outside it, who evidently believe that Kuhn or Quine or Feyerabend has discredited the traditional picture of scientific knowledge. More than that; in this new "post-positivist" era, many scientists (especially social scientists), literati, and philosophers outside of philosophy of science proper have come to believe that the epistemic analysis of science since the 1960s provides potent ammunition for a general assault on the idea that science represents a reliable or superior form of knowing.

Many of my fellow theorists of science, seeing how outsiders have misconstrued our discipline, are persuaded that epistemic relativism is just one of those episodic cultural sillinesses that will wither and die of its own accord. They seem to think that if one either ignores the Kuhns and Feyerabends, or dismisses them with a quick reductio, it will not be long before the situation puts itself right. But, more than a quarter century after the first salvoes from the new wave hit the presses, relativism—about knowledge in general and science in particular—shows no signs of abating. Quite the contrary, the wider intellectual community comes increasingly to suppose that the claims of science to knowledge of the world, even fallibly construed knowledge, have been discredited or at least put in serious doubt. In case my observations about the rampant character of relativism may strike some readers as an exaggeration, I shall quote from a blurb sitting before me, publicizing a conference held in October 1989 at Gustavus Adolphus College, a Lutheran liberal-arts college in the American upper Midwest. The theme of the conference was "The End of Science?"; ironically, the conference itself was officially sponsored by the Alfred Nobel Foundation—the same one that awards the prizes for scientific achievement. The conference announcement opens with this statement:

> As we study our world today, there is an uneasy feeling that we have come to the end of science, that science, as a unified, universal, objective endeavor, is over. . . . We have begun to think of science as a more subjective and relativistic project, operating out of social attitudes and ideologies—Marxism and feminism, for example.

Sober Lutherans have not spoken about science in such terms since Luther's point man, Melanchthon, tore into Copernicus in the sixteenth century. I do not know who the intended "we" is supposed to be in this passage; it certainly does not speak for most philosophers of science. In the face of claims of this sort (and this conference was, I fear, typical of what passes for "the humanities" these days), what is needed—or so it seems to me—is a careful analysis in nontechnical terms of what current work in philosophy of science does and does not permit us to say concerning the nature and limits of scientific knowledge. The issues need to be couched in language that makes them accessible to those outside of the philosophy of science proper but that still comes close to doing justice to the intricacy and structure of the arguments.

However, I did not write this work merely with the aim of setting the exegetical record straight. My larger target is those contemporaries who—in repeated acts of wish-fulfillment—have appropriated conclusions from the philosophy of science and put them to work in aid of a variety of social cum political causes for which those conclusions are ill adapted. Feminists, religious apologists (including "creation scientists"), counter-culturalists, neoconservatives, and a host of other curious fellow-travelers have claimed to find crucial grist for their mills in, for instance, the avowed incommensurability and under-determination of scientific theories. The displacement of the idea that facts and evidence matter by the idea that everything boils down to subjective interests and perspectives is—second only to American political campaigns—the most prominent and pernicious manifestation of anti-intellectualism in our time. The purpose of this short volume is to explore whether the epistemology of science provides—as it is often alleged to—a grounding for such ideological laissez-faire. It is intended both as a purgative and as a prophylactic. A purgative, for those who have already succumbed to the wiles of relativism, mistakenly believing it to be a philosophically coherent position; a prophylactic, for those who—without having jumped one way or another—find themselves perplexed by the claims and counterclaims in the debate between relativism and its critics.

The dialogue form seems ready-made for such situations. When I began writing this book, it was literally a *dia*logue, with precisely two interlocutors—one speaking for the prevailing wisdom in philosophy of science and the other speaking for epistemic relativism. It quickly became obvious that such a two-way conversation could only mislead; for what has become clear since the early 1960s is that there is not one generic position within the philosophy of science but three or four. What all these positions share is a conviction that strong relativism fails to be convincing—even if they arrive at that conclusion from quite different premises. It seemed to me that in order to capture the complexity of that dialectic, I needed a dialogue between a positivist, a realist, a relativist, and a pragmatist. Each of the first three of these characters is a *composite*. I daresay that there is no living philosopher who holds all the views I put in the mouths of my realist, relativist, and positivist. (By contrast, there is at least one person who hews to the line I have my pragmatist defending.) But I have gone to some pains to make sure that the general positions I attribute to the representative from each camp are actually espoused by one or another contemporary philosopher who sails under the appropriate flag.

The hardest task, at least the one I have labored over the longest, is that of finding a suitable voice for the relativist. I believe the relativist position to be profoundly wrong-headed; because I know that about myself and because I am not interested in cheap victories, I have tried my best to make this relativist clever and argumentatively adept. (I note in passing that I have not been helped very much in this task by the sorry state of the relativist literature.) Notwithstanding, I have cited chapter and verse for all the major doctrines that I have my relativist espousing and defending.[1] Those sympathetic to relativism will—*if* the dialogue is successful—disclaim being *that* sort of relativist. Well and good. Nothing would please me more than the discovery that no one is prepared to lay claim to the relativist heritage in this form. But against those who may think I have misconstrued that heritage, I am prepared to defend the claim that much of the literature of contemporary relativism is committed to most of the views under discussion here.

Finally, I am happy to record warm thanks to Philip Kitcher, Deborah Mayo, Cassandra Pinnick, and Adolf Grünbaum, the participants in my 1989 NEH Summer Seminar in Naturalistic Epistemology, and to numerous graduate students for helping me sharpen the ideas for this book. Thanks as well go to the good offices of the National Science Foundation, which made possible research on some of the issues dealt with here.

1. I should note for the record that two of the thinkers who loom large in the conceptual universe of the relativist of my dialogue (Kuhn and Quine) disavow the relativist label. Both are serious and conscientious scholars, who can readily see many of the paradoxes of relativism; but intentions are less important here than consequences. Kuhn's and Quine's writings have unmistakable relativist implications, a fact that few card-carrying relativists have overlooked. One simply cannot address contemporary relativism and ignore how central Kuhnian and Quinean themes have become for that tradition.

Note to the Reader

At a business meeting of the American Philosophical Congress in December 1988, a resolution was passed calling for a committee to be empaneled and charged with preparing a report to the association on "The Current Status of Epistemic Relativism, Especially with Respect to Scientific Knowledge." Accordingly, a committee—consisting of four prominent members of that society with divergent perspectives—was duly appointed. The members were Quincy Rortabender (relativist and author of *Knowledge as Myth: The Outlines of Ethno-Deconstructivism,* and *Skepticism about Everything Except the Social Sciences: A Post-Modernist Guide*), Percy Lauwey (pragmatist; author, *Tinkering with Truth* and *How to Fix Broken Ideas*), Rudy Reichfeigl (positivist; author, *Empirical Adequacy: Who Could Ask for Anything More?* and *Everyman's History of Philosophy: Great Thinkers from Frege to Carnap*), and Karl Selnam (realist; author, *Telling It Like It Is* and *The Irenic Guide to Joint-Carving*). Its members met together over a period of three days during the summer of 1989. Unable to reach consensus (to the evident delight of the relativist on the panel), the committee failed to submit a formal report. However, their sessions were recorded on tape; what follows is a slightly edited transcript of those discussions.

1

Progress and Cumulativity

DAY 1, MORNING

Pragmatist: Gentlemen, I think we should begin, since we are already a bit behind schedule. Having been named chair of this committee, I should say that I interpret my charge to be that of seeing to it that our discussions remain focused on our central tasks and that we do not chase after too many wild hares. We already know one another, having crossed swords on several previous occasions, so I think that no preliminaries of that sort are called for. But we probably should give some thought to selecting the key topics that will form our agenda.

Relativist: Since our brief is relativism, especially as regards scientific knowledge, and since I am the only card-carrying relativist here, I have some suggestions to make about what the salient issues should be. Above all, I think that we should start with the collapse of positivism and foundationalism and move from there . . .

Realist: Forgive me for interrupting, Quincy, but the fact that you are keen on relativism gives you no special claim to set our agenda. All of us here have thought about relativism for a long while; the fact that we have rejected it, and that you have accepted it, is neither here nor there.

Positivist: I wonder if, rather than attempting to legislate our full agenda here and now, we couldn't avoid this procedural wrangling by simply agreeing to start somewhere and then take up the topics as they flow naturally from the exchange?

Pragmatist: I wholeheartedly concur, Rudy. Why don't you suggest a place for us to begin.

Positivist: Well, as we all know, one of the key issues in the epistemology of science has concerned the question of the *growth* of scientific knowledge; thinkers from Peirce to Popper have insisted on the centrality of that problem to scientific epistemology. Nor is it philosophers alone who are preoccupied by it. Scientists and laymen similarly agree that one of the striking features of the diachronic development of science is the *progress* that it exhibits. The philosophical challenge is to find ways to characterize that "progress" as clearly and as unambiguously as we can. Perhaps therefore this would be an appropriate place for us to begin our explorations. And since our brief is to examine the status of contemporary relativism, maybe we could ask Quincy to kick off the discussion with a characterization of the relativist view of cognitive progress.

Relativist: I have no objections to our starting there, if you like, since I share your view that people have an abiding faith in the progress of science. Equally, however, I want to go on record straightaway as having grave reservations as to whether there is any robust, objective notion of the growth of knowledge. I happily grant you that our theoretical understandings and representations of the natural world change dramatically through time, though whether those changes represent "progress" or simply change is unclear. But I think that it would be reversing the natural logic of this subject for me to start things off. Most of us relativists reject the notion of progress because the two well-known accounts of scientific progress—associated with positivism and realism respectively—have been dismal failures. Accordingly, and I can assure you not out of any shyness on my part, I would urge Rudy or Karl to tell us whether they have a coherent theory of scientific progress to put forward. I will fill out my position in response to what they have to tell us.

Pragmatist: That's certainly agreeable as far as I'm concerned. Perhaps we can call on Rudy to outline the problem of progress or theory change as he sees it, since the positivist account of scientific progress probably remains the best known.

Positivist: Gladly. In a nutshell: science is the attempt to codify and anticipate experience. The raw materials of science are observational data or measurements. We develop theories and laws to correlate, explain, and predict those data. A science *progresses* just to the extent that later theories in a domain can predict and explain more phenomena than their predecessors did. Since the seventeenth century, the sciences—at least the natural sciences—have done just that.

Relativist: Hang on a minute. When you talk about what a theory "can" predict and explain, are you referring to what it *has* explained and predicted or are you talking about everything that it might be able to predict and explain?

Positivist: You can take it in either sense since science exhibits impressive credentials of both sorts.

Relativist: Well, if we focus on the first sense, what one might call demonstrated progress, I will grant you that some theories have managed to predict and explain some things not guessed at by their rivals. But I'm not sure that this difference gives us a sound basis for maintaining that one theory is really better or truer than another. After all, the fact that one theory has more proven predictive successes to its credit than a rival might be just an artifact of how long each has been around, how assiduously their applications have been explored, how many scientists have worked on them, etc. You surely don't want to argue that the goodness of a theory is a matter of such accidental circumstances as these?

Positivist: Well, as I said earlier, genuinely progressive theories are those which have the capacity to explain and predict a larger range of facts than their rivals. That is, in part, a *prospective* matter. I accept your point that one theory's known successes might have more to do with these accidents of history than with anything about the theory itself.

Relativist: But if you're saying that, in deciding whether one theory represents progress over another, we have to compare their *prospective* explanatory and predictive ranges, then I don't see how we could ever settle that issue since—as you just noted—we can never know all the consequences of any theory, infinite as that set is. You face a dilemma, Rudy: comparing the known achievements of rival theories can be done but is indecisive since those achievements will be in part a function of various accidents of the distribution of labor in the scientific community; yet it is impossible to compare the *potential* but unknown capacities of rival theories.

Positivist: Quincy is right, up to a point. A theory, any theory, has an infinite range of consequences only some of which will have been examined at any chosen stage of inquiry. But we are nonetheless often in a position to make dependable judgments about the prospective scope of rival theories, even when (as will always be the case) we have actually tested only some of those theories' consequences. Indeed, I can describe for you a procedure which will allow us to make such judgments *even if we have never tested any of the rivals.* Let us suppose that we have two theories under consideration, call them T_1 and T_2. Suppose further that we can show that T_2 *entails* T_1. Under such circumstances, we know that T_2 must have all the consequences of T_1, as well as some additional consequences besides (provided that T_1 does not entail T_2). Hence if a later theory entails an earlier one, but is not entailed by it, then we know that the later theory must be more general than the earlier.

Relativist: I'm not sure I see the point . . .

Positivist: It's simply this: A few moments ago, you said that the comparison of the prospective successes of rival theories was impossible, suggesting thereby that we positivists have no viable theory of scientific progress. What my latest example shows is that we can often demonstrate, even prior to *any* testing, that one theory is more general than another.

Pragmatist: But surely generality alone, in the sense of a maximally large class of entailments, is not scientific progress. If it were, then the adumbration of tautologies—which imply all true statements—and contradictions, which entail everything, would represent the ideal end point of science.

Positivist: What's wrong with tautologies and contradictions alike is not their lack of generality but their nonamenability to empirical test and thus their low information content. When I say that one theory represents progress over another as long as the former entails the latter (but not vice versa), I mean to refer only to theories per se, i.e., to sets of universal statements which are genuinely empirical by virtue of their prohibiting certain states of affairs. Tautologies and contradictions prohibit nothing and are thus not in the class of theories.

Relativist: What you appear to be saying is that scientific progress can occur only if (a) one *testable* theory succeeds another and if (b) the later theory entails the earlier. But Duhem, Quine, and a host of others have shown that scientific theories are not falsifiable and hence not testable.[1] They showed specifically that any theory whatever can be retained in the face of recalcitrant evidence, provided we are prepared to make drastic enough changes elsewhere in our framework of beliefs.

Pragmatist: I think, Quincy, we should tackle one issue at a time. We all know you believe theories are nonfalsifiable in principle, and I for one am prepared to set aside one of our later sessions to deal specifically with that issue. But I wonder if for now we shouldn't allow Rudy to finish setting out his position on scientific change and progress.

Positivist: Thanks for the intervention, Percy. Our relativist friend was overhasty, for I should be the last to claim that greater generality in our theories is a *sufficient* condition for scientific progress. My claim thus far was simply that greater generality was a *necessary* condition for making a well-founded claim of progress.

Pragmatist: Well, what more is required?

Relativist: Rudy is doubtless on the verge of answering that question, but I wonder if I could interrupt here. He just told us that greater generality was a necessary condition for scientific progress, and I have no quibbles with that claim as far as it goes. But I think we have been too hasty in accepting Rudy's assimilation of greater generality to some logical relation of entailment.

Positivist: You're unclear about what an entailment is?

Relativist: In fact I am, since it seems to me that what a theory

1. For most of the classic texts on this problem, see Harding (1974).

entails depends upon the other theories and assumptions with which it is conjoined; but that is not my worry for the moment. My problem is this: I grant you that if one theory entails another theory and not vice versa, then the former is more general than the latter. But surely we can think of situations in which one theory is more general than another even when the relevant entailment relations fail to obtain.

Positivist: Such as what?

Relativist: Well, I'm inclined to think that the theories of quantum mechanics are more general than theories of ecology, even though I couldn't begin to derive the latter from the former.

Positivist: You're surely right, but I don't think I'm committed to claiming that one theory is more general than another only if an entailment relation obtains between them. I see one-way entailments as a sufficient condition for making judgments of generality, not a necessary condition.

Relativist: So, on your view, one theory might be more general than another, and thus potentially progressive, even though neither theory entailed the other?

Positivist: Of course; why are you bothered?

Relativist: It seemed to me that you were suggesting that deductive logic was enough to settle issues of progress, and I have the impression that it is otherwise.

Pragmatist: I wonder if we couldn't return to where we were before this digression. I think that Rudy had said that progress judgments require more than greater generality. What else is involved?

Positivist: Well, above all, we expect the more progressive theory to be *better confirmed* than its predecessor. We expect it to have enabled us to explain and predict phenomena which its predecessor either couldn't explain at all or which its predecessor predicted incorrectly.

Realist: But if the earlier theory made an incorrect prediction and the later theory—by virtue of entailing its predecessor—and I believe you called that a necessary condition for progress—exhibits all the consequences of its predecessor, then any false prediction made by an earlier theory is also going to be made by its successor. Hence how could a later theory possibly *both* entail its predecessor *and* predict something correctly which "its predecessor predicted incorrectly"?

Pragmatist: Are you suggesting, Karl, that the old Baconian ideal that later theories should "contain" their predecessors is bankrupt because it would require later theories to incorporate all the failures of earlier ones?

Realist: Exactly. Once we realize that most theories in science are given up precisely because we have found them to be false, then it follows that the last thing we want to insist on is that their successors must capture all their empirical consequences! Perhaps I can put my challenge to the positivists most concisely in this form: I can see that a later and more general theory might well make some predictions—including correct predictions—concerning matters about which its predecessor was wholly silent; but how could the later theory manage to avoid the incorrect predictions made by its predecessor if it entails that predecessor?

Positivist: You have a point, Karl. Answering it will require me to be a bit more precise and detailed than I have been thus far. Let us distinguish, within the context of what I have been loosely calling a theory, between two elements: the theory per se and the associated experimental laws. Laws coordinate observations, and theories coordinate laws. Now, when we discover that a theory has broken down, that is has some false consequences, what we are really discovering is that some of the lawlike generalizations coordinated by that theory are false, i.e., they are not laws at all. Of course, as Karl says, we don't want to demand that a later theory must replicate the known failures of its predecessors.

Relativist: So now what is your story about the relation between successive theories in a progressive science?

Positivist: I suggest that what we expect a progressive theory change to do is to produce a successor theory which (a) retains all the *nondiscredited,* lawlike statements associated with the earlier theory, (b) drops out those pseudo-laws which have already been refuted, and (c) introduces some new lawlike regularities not previously encompassed within the predecessor theory. Things are even clearer if (d) some of the lawlike statements associated with the later theory, and not embraced by its predecessor, correctly predict hitherto unexplained and unpredicted phenomena. When all these conditions obtain, then we have a paradigmatic case of progressive theory change.

Realist: Although my positivist friend and I disagree about many matters, I find myself almost wholly in agreement with the characterization he has just offered. Perhaps consensus on these issues is within our reach.

Pragmatist: Let's not be too hasty. Rudy's definition of theoretical progress in science does beg a few questions. For one thing it requires us to accept a distinction between a theory per se and the lawlike statements associated with it. As I understand it, that distinction is to be drawn chiefly in light of a distinction between observational terms—which is what occurs in the "laws"—and nonobservational terms which occur in the theory per se. Have I got that right?

Positivist: Of course.

Pragmatist: In that case, I cannot accept your distinction, since I doubt that there is a sharp line between theoretical terms and observational terms; and I suspect that Quincy will find it equally objectionable. And so should Karl, despite his initial willingness to accept Rudy's account of progress, since Karl's realism about science hinges crucially—if I understand it—on the repudiation of any sharp observational/ theoretical dichotomy.[2] But even if I were to grant you, Rudy, that there is a viable distinction between what is observational and what is not, I would still have problems with your characterization of scientific progress.

Positivist: And what would those be?

2. Most realists (e.g., Maxwell, Sellars, Popper, and Putnam) hold that the collapse of the theory/observation dichotomy undermines instrumentalism and paves the way for realism about theoretical entities. It should be noted that this way of putting the point smacks of the paradoxical since a realism "about theoretical entities" would seem to presuppose the very distinction whose avowed demise it celebrates.

Pragmatist: You originally told us that a sine qua non for progress was that later theories must entail their predecessors. When some of us noted that such a policy would involve retaining all the failures as well as the successes of earlier theories, you backed off and conceded that entailment between theories was too strong a condition. Now you are telling us that one theory is better than another if, among other things, the later theory retains all the nondiscredited lawlike statements of its predecessor.

Positivist: Quite.

Pragmatist: But how does one tell what those are? Since no one believes that we are fully aware of all the empirical or observational consequences of any theory, how can we ever be reasonably confident that a later theory has retained all the (correct) observational consequences of its predecessor?

Relativist: That sounds a bit like my earlier worry about how we can possibly compare the unknown, prospective features of different theories.

Pragmatist: Indeed it does. For all we know, a new theory might have ignored or dropped out many of the correct but unknown laws associated with an earlier theory. Thus, giving up the earlier theory, because of certain *known* failures, and replacing it by a theory which incorporates the earlier theory's *known* successes and avoids its *known* failures—if we could manage to do that—offers us no assurance that the new theory will generally work better than the old one might have.

Positivist: Your point is well taken, but there is an answer to your worries. Specifically, I require that, for progress to occur, successor theories must capture their predecessors as "limiting cases." This requirement enables us both to retain the successes of the earlier theory and to correct for its mistakes. Indeed, this is precisely what happened in physics at the turn of the century when Einstein was able to show that classical mechanics was a limiting case of relativity theory and when Planck and Bohr were able to show that classical electrodynamics was a limiting case of quantum theory.

Realist: Rudy, there is some sleight of hand going on here. You were originally trying to convince us that successive theories in a progressive science *entailed* their predecessors, while going beyond them. You have now shifted to telling us that progressive successor theories must capture their predecessors as "limiting cases." But am I not right in thinking that if one theory, T_1, is a limiting case of another, T_2, then T_1 may have some consequences not exhibited by T_2?

Positivist: Naturally, and that's a virtue of this analysis; it allows a later theory to capture the relevant successes of its predecessors while avoiding some or all of its failures.

Realist: But what you gain on the swings you lose on the roundabouts. For if T_2 does not entail T_1 but merely captures some of its associated laws as limiting cases, then how can we be confident that all of T_1's *true but unknown* consequences are captured by T_2? Isn't it at least conceivable that classical mechanics, say, had some true consequences which are not explained or predicted by relativity theory?

Positivist: What you say has some surface plausibility, I suppose. But if we are prepared to think, following Hertz, that a theory is just a system of equations, then provided we can show that the equations constituting an earlier theory are derivable—at least as limiting cases—from the equations of a later theory, then we surely have strong grounds for arguing that the later theory will enjoy all the successes of the earlier.

Realist: Even if I were to grant that the equations of classical mechanics are limiting cases of the equations of relativity theory, I can scarcely go along with your naive identification of a theory with the equations with which we represent some of its most fundamental laws. Theories are, after all, much more than sets of equations; they typically involve claims about the basic causal processes and fundamental entities in a domain. Thus, Newton's mechanics is not only his three laws of motion, plus the law of gravity; equally, Newtonian "theory" is a complex set of concepts about absolute space and time, about the nature of matter, forces, and the like. I'm afraid, Rudy, that you cannot begin to grasp what scientific change is about until you realize that successive scientific theories are complex networks of assumptions about the basic building blocks of the world and about how they interact.

Positivist: I'm not sure I see what you're driving at.

Realist: What I'm saying is that even if you establish that the equations associated with successive theories are homologous (and that, at best, is what limiting-case relations exhibit) you are still far from having established that successive theories approximate to one another in the terms that really matter, i.e., in their underlying mechanisms and theoretical entities. No one has ever shown that Newton's mechanics—as opposed to a few mathematical laws associated with it—is a limiting case of either the special or the general theory of relativity.

Positivist: The fact that no one has shown this in its full generality does not mean that it is not so.

Realist: That much is true. But I can give you a perfectly general proof that Newton's theory in the full-bodied sense could never be a limiting case of the general theory of relativity—or of any other theory for that matter.

Positivist: I'm not clear how relevant that would be to the issues at hand, but I'd nonetheless like to hear the argument.

Realist: It's quite simple. By definition, limiting-case relations can be established only between sets of equations. That means that any theoretical claims which are strictly "qualitative" rather than "quantitative" in character can never be a limiting-case of other claims. Newton's physics is full of such qualitative claims. For instance, he asserts that light is corpuscular in character; that, indeed, is the core assumption of his optical theory. But what is the equational representation of that hypothesis? Again, he asserts that repulsive forces of some sort are responsible for the phenomenon of surface tension. Notoriously, Newton holds that space and time are absolute. How, without more detail than Newton provides, is one to represent that in a form which would lend itself to limiting-case relations?

Positivist: I don't think that you and I, Karl, will ever agree on the nature of theories and their logical structure. Those assertions to which you are referring are not part of what I mean by Newtonian mechanics. They are perhaps a part of Newton's speculative philosophy of nature but not of his physics. Yet I daresay that to defend my views on this topic would take us too far afield. After all, we're supposed to be exploring the status of relativism, and instead positivism is becoming everyone's target.

Pragmatist: The issue, I think, is whether positivism provides a coherent picture of scientific progress. If it doesn't, then we must at least concede that the relativist is right about this much: the philosophy of science that has long been dominant offers no coherent account of scientific progress. And this is important because I seem to remember that Quincy earlier claimed that much of the rationale for his position derives from the collapse of the position which you have been attempting to defend.

Positivist: I am willing to entertain objections from relativists like Quincy, but I get irritated when realists like Karl suggest that I have no coherent account of scientific change. After all, it is the realist's construal of theories, seeing them as bloated ontological instruments, that sets up the grounds for the relativist's critique of scientific knowledge.

Realist: How can you say that? Realism is the only epistemology of science that provides a cogent alternative to relativism about science.

Positivist: What I mean is that the realist view of scientific theories—a view which sees them as making a range of claims about the world which go far beyond anything which can possibly be observed or directly tested—invites the relativist riposte that theories, if construed in that fashion, are articles of metaphysical faith rather than claims closely tied to the evidence. Indeed, if we accept the realist picture of scientific theories, then we are quickly led down the garden path to incommensurability, to the indeterminacy of translation, and ultimately to the inscrutability of reference.

Realist: That's absurd!

Positivist: Has it never occurred to you, Karl, that Feyerabend and Kuhn, two of the best known contemporary relativists, became relativists precisely because they construed rival theories as metaphysically pregnant "worldviews," which made a host of claims about matters beyond any conceivable empirical adjudication?

Pragmatist: You may be right, Rudy, about the realist analysis of science providing ammunition for relativism; but in this context your attack on realism is purely diversionary, for you have yet to show that positivism has a theory of scientific growth or progress which insures that later theories are really better than earlier ones. And without that, you've set the stage for relativism yourself. So, if I may, I'd like to bring you back to focus on that point.

Positivist: I'll give it one more try. As I reflect on the various worries that you three have been raising, the central one seems to be this: most of my characterizations of scientific progress have been concerned to give us ways to compare the thus-far unexplored potential of rival theories. In various ways, you have repeatedly argued that there is a problem about comparing the prospective features of rival theories. I am persuaded by the arguments raised that a theory of scientific progress which involves a priori projections about the prospective success of rival theories is not manageable. Accordingly, I shall propose that we conceive of scientific progress entirely in terms of the *demonstrated successes* of rival theories.

Realist: Can you be more precise?

Positivist: I was about to be! Let me reformulate my demands for progress in this fashion: for scientific progress to occur, (a) a successor theory must embrace all the *confirmed* true consequences of its predecessors. By putting it so, I hope that Karl can now see why I suspected that his earlier objection to limiting-case relations was a red herring.

Realist: I can see that if all a later theory need preserve of its predecessor are the latter's confirmed consequences, then our discussion of Newton's "qualitative" beliefs is not really germane since few of those beliefs generate what we could regard as confirmed consequences.

Positivist: A second condition for scientific progress is (b) that the successor theory must also exhibit some empirical strengths not shown by its predecessor.[3] Of course, this is no ironclad guarantee that the new theory will always hold up better than its predecessor would have, since we can conceive of circumstances in which the untested consequences of the older theory might stand up better than the untested consequences of the newer theory. But there *are* no guarantees where empirical research is concerned, and it is a skeptic's fallacy to demand them. We can judge the future only by the past. If we can show that one theory enjoys all the known strengths of its predecessors, that it avoids some of its predecessor's mistakes, and that, besides, it can explain some things not explained by its predecessor, then we have very powerful reasons for regarding the new theory as an improvement over the old. That is what scientific progress amounts to. And, having offered you that definition, I will go on to make the claim that the natural sciences—and only the natural sciences—exhibit progress of that sort.

Realist: I'm curious why you don't add an obvious third condition: (c) that the later, more progressive theory must avoid some of the false consequences of its predecessor.

Positivist: When that occurs, we have yet further evidence for the progressiveness of the successor theory. But I did not want to make your third condition (c) a necessary one for progress since that would force us to say that one theory could never be progressive over another unless the latter was known to have some false consequences.

Realist: And what's wrong with that?

Positivist: Simply this: I want to leave open the possibility that we may judge one theory to be an improvement over another, even when neither has *yet* been falsified.

Realist: That seems reasonable enough.

Relativist: So far, I have been content to let others do my work for me, but I must strenuously object to this latest redaction of your position, Rudy. I object because it—like every one of your other formulations—depends upon a degree of *cumulativity* during theory change which is wholly belied by

3. This position has been developed by Post (1971), among others.

the historical record. You began by telling us that later theories must entail their predecessors. They don't. You then told us that earlier theories in a science turn out to be limiting cases of later theories. Historical research of the last three decades makes clear that this condition is rarely, if ever, satisfied. Finally, you now tell us that later theories in a progressive science must preserve all the known empirical successes of their predecessors. Again, they don't.

Positivist: Why do you say that?

Relativist: Kuhn and Feyerabend showed repeatedly that there are explanatory losses as well as gains in most theory transitions; because that is so, an account of progress which requires—as yours does—the cumulative retention of known empirical successes from one theory to another is simply demanding too much. If progress is understood as you just defined it, then we have no grounds whatever for holding natural science to be a generally progressive activity. I'll go further than that. In my view, there is no viable notion of cognitive progress which does not require a high degree of cumulativity between earlier and later theories in a science. Rudy was surely right to look for some such explication of progress. The unhappy fact of the matter, however, is that successive theories in the sciences—including the most mature and well-developed sciences like physics—fail to exhibit the requisite degree of cumulativity. That is one reason why a theory of scientific progress is an absolute nonstarter. In a word: no cumulativity, no progress.

Positivist: All this twaddle about recurrent losses of established empirical successes is just hogwash. I defy you to give me an example of a major theory change in the mature sciences which involved such losses.

Relativist: What counts for you as a mature science? Physics since Galileo?

Positivist: Fine. You have my prejudices down pat!

Relativist: Well, what about a case like this: The vortex physics of Descartes explained why the planets moved in the same direction and the same plane, namely, because they were all carried around the sun by a vortex.

Positivist: And so?

Relativist: Well, Newton nowhere in the *Principia* explains those salient facts about planetary motion.

Realist: But Newtonian mechanics does not forbid that the planets might move in the same direction and in the same plane.

Relativist: Right. But my point is that Newton had no explanation for these phenomena. Compatibly with Newtonian celestial mechanics, the alternate planets could have moved in opposite directions and in planes perpendicular to one another. Newton was forced to *assume,* as an initial condition of the solar system, that the planets moved in these ways; whereas Descartes could *explain* these same phenomena. What was an empirical success for the earlier theory was an unexplained fact for the successor. That's what I mean by loss in the history of science; and there are loads of other cases.

Positivist: I think you're being altogether too hasty here, Quincy. After all, Kant and Laplace both showed how—on Newtonian principles—one could get the solar system configured as it is. The nebular hypothesis offers a Newtonian explanation for these facts.

Relativist: So it does; but need I remind you that the nebular hypothesis emerged a century *later* than Newton's *Principia*—and long after virtually all physicists had become Newtonians? If you want to show that the history of science is rational, you must show that there were good reasons at the time for the acceptance of Newtonian mechanics. On your view, we should have to say that it was reasonable to believe that Newtonian mechanics was an improvement on Cartesian physics only by the end of the eighteenth century, fully five decades after everyone doing physics had accepted Newton.

Positivist: I still think my general point is right; for here we have a situation where the later theory had the resources to capture all the known successes of its predecessor, even if physicists at the time didn't know it. The fact remains that the later theory did cumulatively retain all the known successes of the earlier one.

Pragmatist: You and Rudy have reached a bit of an impasse on that issue, so—by way of keeping the conversation moving—I would like to challenge an assumption that you both brought to the discussion. Specifically, you have supposed that cumulative theory change in some sense or other is a precondition for making judgments of scientific progress. I see no reason why the only sorts of changes we call progressive or contributory to the growth of knowledge need be transitions which retain cumulativity. I suspect that this claim will make me odd man out in this discussion, since our realist friend Karl, every bit as much as Rudy and Quincy, takes cumulative retention of empirical success to be a sine qua non for scientific progress. Am I right?

Realist: Of course I do, despite my quibbles with Rudy about whether limiting-case talk is the best way to capture what those retentions are. As such realists as Boyd and Putnam have argued, in any mature and well-developed science later theories entail at least approximations to their predecessors.[4] As a realist, I hold that science through time is moving closer and closer to a correct characterization of the natural world. Because that is so, later theories need to preserve the known successes of their predecessors. If they did not, there would be no coherent sense that we could attach to the notion of science progressively approximating to a true account of the world. Rudy and I are of one mind on this issue, even if we disagree about the nature of theories.

Pragmatist: I'm afraid that I stand with Quincy where the historical record is concerned. Typically, later theories do not entail their predecessors, nor capture them as limiting cases, nor retain in wholesale fashion all their known empirical consequences.[5] But unlike the rest of you, including Quincy, I do not see, in that failing, any grounds for pessimism about the possibility of developing a theory of scientific progress.

Relativist: But scientific progress without the cumulative retention of our successes is no progress at all. I think it was Kuhn who pointed out that, if there are losses as well as gains

4. See Boyd (1973) and Putnam (1978).

5. For a detailed development of this argument, see Laudan (1984, chap. 6).

associated with theory transitions, then there can be no objective way of telling whether the gains outweigh the losses.[6] You have conceded, Percy, that such losses occur yet you continue to hold that science is progressive. It just won't wash.

Pragmatist: Our notion of progress carries with it a lot of baggage, I'll grant you. But neither in science nor elsewhere need we make progress dependent on some sort of total cumulativity. What, after all, is progress—whether scientific or otherwise? We judge an activity to be making progress when it is further along toward the realization of its ends now than it formerly was. Progress is thus a diachronic notion that involves reference to an aim, or set of aims, and an empirically based ranking of the degree to which various efforts at the realization of those aims have in fact furthered them.

Positivist: Are you suggesting that scientific progress is no different from progress in any other area?

Pragmatist: Yes and no. "Progress" is a perfectly general notion of successive movement towards the realization of an end; to that extent progress in science is like progress in, say, bank robbing or arms-control talks. But to the extent that science has a unique set of aims (and I suspect that we will eventually have to fight out that issue in these conversations), then progress towards the realization of scientific ends may be different from other forms of progress just because the ends are different. But in all these cases, progress is movement towards the realization of one's ends.

Realist: I have no objection, Percy, to your defining progress in science instrumentally, as a matter of attempting to realize certain ends, but I can't see how that circumvents the cumulativity issues we were discussing before. Whatever else the ends of science might be, they surely include things like "explaining and predicting everything that happens in the natural world." If so, and if losses of the sort that you and Quincy are so insistent about actually occur, then we are in

6. Kuhn makes this point at several places in his work. See, especially, Kuhn (1970, pp. 101–2, 108, 110, 111, 118–19).

no position to say that later theories do more to further our ends than earlier ones did.

Pragmatist: I see it rather differently, Karl. Suppose we were to say that one of the central aims of science—perhaps *the* central aim—is to produce theories which are increasingly reliable. Suppose, further, that we unpack that notion of reliability in terms of the ability of theories to stand up to more and more demanding empirical tests. Suppose, finally, that we make certain plausible assumptions about what counts as demanding tests. We might say, for instance, as realists like you are prone to, that theories which make surprising predictions successfully have been more robustly tested than those which do not. Equally, we might say that theories which pass tests by means of controlled experiments have been more convincingly tested than those which are merely tested against haphazardly collected observations. Again, we might say that a theory which has been tested in several domains of its application has been better tested than one which has been tested in only one. Now, to a crude first-order of approximation, my theory of empirical progress simply says that one theory represents progress over another provided that the one has passed tests of a sort that the other has failed to pass (whereas the latter has passed no sorts of tests which the former has failed). As long as our knowledge is becoming more reliable, progress is being made.

Positivist: It's all very well, Percy, to say that in a progressive science later theories will have been better tested than earlier ones; but I'm afraid that it's to no avail. For if explanatory losses occur during a theory transition (and you and Quincy are emphatic that they do), then we have to acknowledge that the older theory passed certain tests which its successor failed to pass. We are back to square one. Progress requires cumulativity.

Pragmatist: You are too hasty, Rudy. Quincy and I have argued that earlier theories typically entail or even *explain* some things not explained by their successors. But—and here Quincy and I part company—that is *not* the same thing as saying that the predecessor theories have *passed certain tests* failed by their successors, let alone saying that the predecessors passed *kinds of tests* different from those passed by their successors.

Positivist: Do explain, since I found what you just said wholly unintelligible.

Pragmatist: Gladly. There are two points really. We need to distinguish, first, between tests and nontests of a theory, and then, among the tests of a theory, between those which are more severe and those which are less severe. Once such distinctions are drawn, I think it will be clear why lack of cumulativity is no particular threat to the progressiveness of the scientific enterprise. First, then, tests and nontests. Suppose that I find some puzzling fact that piques my curiosity; maybe I find a massive fossilized bone while digging in my backyard. I may develop a low-grade "theory" to explain this fact: perhaps I conjecture that God put it there to test my faith in the literal reading of Scripture. Now although my hypothesis arguably explains the fact of a fossil bone in my backyard, that hypothesis is not *tested* by the fossil bone. On the contrary, my hypothesis was specifically constructed to explain the bone. My hypothesis might be testable but I would have to look further afield to find something which counted as a genuine test of it.

Relativist: I am unclear as to what notion of 'test' you are working with. You seem to regard the concept as clear and straightforward where it seems to me a very problematic notion.

Pragmatist: One could, indeed people do, write whole books about what a "test" is, so I will not pretend that this is an unproblematic notion. But this much seems both clear and uncontested: an observation or set of observations is a "test" of a theory only if the theory or hypothesis might conceivably *fail* to pass muster in light of the observations. If, as in my hypothetical case, the theory was invented specifically to explain the phenomenon in question, and was groomed specifically so as to yield the result in question, then there is no way in which it could fail to account for it. *Where there's no risk of failure, there is no test involved.*

Relativist: Well for my part I'm not sure that there are *ever* circumstances in which a theory runs a risk of failing, but I will suppress those concerns for now. Do proceed.

Pragmatist: Thanks, I shall. The first point I wanted to make was simply this: *there are some things a theory entails (and even explains) which are nonetheless not tests of the theory in question.* Thus, a theory may explain what it was specifically invented to explain, even though—as I've just said—those phenomena may constitute no test of the theory. Since there is, then, a difference between the phenomena a theory explains and the phenomena by which it is tested, it follows that we should not suppose that simply because an earlier theory explains certain facts not explained by its successor that the earlier theory has thereby passed certain "tests" not passed by the successors.

Positivist: I was quite persuaded by your earlier argument that there is a difference between the phenomena which a theory entails and those which constitute tests for a theory. But I'm a little perplexed about the wedge you are now driving, if I understand it rightly, between things a theory explains and those things which confirm the theory. I've generally tended to think that a theory's confirming instances and the things it explains boil down to the same set.

Pragmatist: Would an example help?

Positivist: Sure, if you've got one.

Pragmatist: See if this will do. As we all learned in high school, Newton's theory synthesizes the laws of Galileo on free fall and Kepler's laws for planetary motions. Let me ask you a couple of questions about this case. Suppose that I go out and make some measurements about the rate of fall of stones in vacua. Imagine that what I find supports Galileo's law. Obviously it also therefore supports Newton's theory of gravitation. Right?

Positivist: Of course.

Pragmatist: Now, Rudy, come my two questions: First, are we agreed that Kepler's laws neither entail nor *explain* the fall of stones?

Positivist: Of course they don't.

Pragmatist: Yet—my second question—doesn't this new evidence about falling stones increase our confidence in Kepler's laws?

Positivist: On the principle that if our confidence in a statement increases, then our confidence in everything entailed by that statement also increases, I guess we must say yes. If we become more confident of Newton's theory by virtue of experiments on falling bodies, then we have to grant that we thereby become more confident about everything that follows from Newton's theory, such as Kepler's laws.

Pragmatist: So, as you have seen for yourself, there are cases where a statement (in this case Kepler's laws) is supported by evidence which that statement neither entails nor explains.

Realist: I don't find this case very convincing. Percy seems to be assuming that I share his "intuition" that new support for Galileo's laws automatically provides support for Kepler's laws via the bridge established between them by Newton's theory of gravitation. But what if Newton's gravitational theory is wrong in classifying free-fall and planetary motions as analogous phenomena? In that case, evidence for Galileo's laws would be no evidence at all for Kepler's laws. I think we should keep it clean by saying that statements are supported only by their entailed consequences.

Positivist: If I remember rightly, that idea is what Hempel once called the "Nicod criterion."[7]

Pragmatist: There are two ways we might formulate that idea, which have very different implications. The weaker version might be this: if a statement *s* entails *e,* and *e* is true, then *e* supports *s.* The stronger version of a Nicod-like criterion would insist that a true *e* supports *s* only if *s* entails *e.*

Positivist: So the weaker version makes the existence of an entailment relation between a theory and an observation a *sufficient* condition for calling the latter evidence for the former. While the stronger version makes entailment a *necessary* and sufficient condition?

Pragmatist: Precisely. I happen to think that both versions are

7. See Hempel (1965, chap. 1).

false. But let me focus now on what I am calling the stronger version. As I understood Karl's argument, he was denying my claim in the Newton-Galileo-Kepler case that support for Galileo's law automatically provides support for Kepler's laws. I gather he did so because he accepts what I am calling the strong version of the Nicod criterion. As he sees it, since Kepler's laws entail nothing about falling stones, those laws can derive no support from evidence about falling stones.

Relativist: Exactly.

Pragmatist: I need to show, if I can, that the strong version of the Nicod criterion is wrong-headed.

Positivist: I quite agree that the burden of proof falls on you, Percy.

Pragmatist: Consider the following example. Suppose that I am contemplating the status of the following claim: "The next crow I observe will be black". Let's call that hypothesis *s*. Now suppose I am given the following information: 10,000 recent sightings of crows have revealed all of them to be black. Let's call that information *e*. Obviously, here *s* does *not* entail *e*. Yet, equally obviously, *e does* provide support for *s*.

Relativist: And just what is this tedious example meant to show?

Pragmatist: It shows that any quick identification of what a statement entails with the instances that lend support to the statement should be stoutly resisted. As the crow case shows, there is evidence which can lend strong support to a statement, even when that evidence is not entailed by the statement in question.

Realist: Hold on a second, Percy. What the argument you've just given us shows is that there are some supporting instances for a statement or theory which are not among the things the theory explains or entails. I can buy that. But I think you are committed to showing something other than this result. When Rudy first raised his doubts concerning your claims about explanation and support, *I* took him to be claiming merely that all the instances which a theory explained or entailed would count among its supporting instances.

Pragmatist: What I earlier called the "weak" version of the Nicod criterion?

Realist: Precisely. In showing us that some supporting instances of a theory fall outside the scope of the theory's empirical consequences namely, that the Nicod criterion in its strong version is false, you have failed to address the question as to whether there are things which a theory entails which fail to support the theory.

Relativist: Karl's got you there!

Pragmatist: Thanks for the clarification. But if that was Rudy's particular worry, it seems to me that we have already answered it.

Realist: How so?

Pragmatist: Well, I thought we had agreed half an hour ago that theories are generally not tested by those phenomena which they were expressly invented to explain. That result is sufficient to establish that some of the empirical consequences of a theory may fail to count among its supporting instances.[8]

Positivist: You're saying, if I understand it rightly, that the explanatory losses in the history of science might not involve loss of genuine supporting instances?

Pragmatist: I'd put it this way: it may turn out that a later theory has passed *all* the tests passed by its predecessors, *even if* there are some phenomena which those predecessors can explain or entail which the successors cannot. Hence loss of explanatory or empirical content per se is not threatening to judgments of scientific progress.

Realist: Are you suggesting that the only losses shown by the history of science are of this sort, what you might call explanatory losses rather than "testing losses"?

Pragmatist: No, I'm not. Such a claim would require more

8. Throughout this discussion, the participants have been making the simplifying assumption that a theory's explanatory content and its empirical consequences are coextensive. It is thought that some of the participants would probably take exception to that identification. But the argument being made here, that a theory's supporting instances are not coextensive with either its empirical consequences or its explained instances, is sound on either construal.

historical scrutiny than I've been able to give it, although there are some prominent cases of just this sort. Recall, for a second, Quincy's example, when he was trying to convince Rudy that theory losses are ubiquitous. He reminded us of the case of Cartesian vortex theory, which was able to give a causal explanation for the codirectionality of planetary motions which Newton was later unable to explain.

Relativist: Surely you're not going to try to show that Newtonian mechanics could handle this case from the beginning?

Pragmatist: No, I'm not. But I will now say to you what Rudy should have said then: Newtonian physics was under no obligation to account for that phenomenon.

Positivist: Why ever not?

Pragmatist: As we all know, Descartes' vortex theory was specifically invented to explain this feature of planetary astronomy. Hence it would have been inappropriate to regard that phenomenon as a test of Cartesian theory, and thus futile to fault Newton's theory for failing to explain it. More generally, what I want to stress is that the instances which test a theory are not necessarily the same as the instances which a theory explains. Since Kuhn and Feyerabend have focused exclusively on explanatory loss, the acknowledgment of such losses entails nothing about there also being losses of genuine "test instances." Even so, I would not go so far as to say that all cases of loss fall under this type.

Realist: So, you concede that there may well be cases where an earlier theory passes some genuine tests not passed by its successor. In that case, you no more have a robust theory of progress than the rest of us do.

Pragmatist: Give me time, Karl. You'll recall that I referred moments ago to a second distinction which provides additional support for my case. There is general agreement among methodologists of science that certain types of tests of theories are more probative than others. For instance, the ability of a theory to stand up to tests in quite diverse domains of application is a better guide to the theory's reliability than its ability to stand up to tests in a single domain. Similarly, a theory's ability to predict surprising phenomena correctly counts for more than its ability to predict wholly expected results.

Relativist: A caveat is in order here. I don't mean to stop you from developing your case, but I do want to note that even if most so-called methodologists agree that certain tests are more probative than others that does not make them so. I shall eventually want you to explain to me why we should regard *any* form of test as probative, not to mention why we should hold some tests as more probative than others.

Pragmatist: Duly noted, Quincy. To continue: Let us suppose that an older theory T_1 has passed a certain test t and that a later theory, T_2, has passed a different and more demanding test t'. Suppose, further, that T_2 simply fails to address the phenomena that generated t. Now—and I'm trying to contemplate the hardest sort of case—we have here a loss which is not only an explanatory loss but also a testing loss.

Realist: Do you mean to say that T_2 *fails* test t?

Pragmatist: No, I do not. If we are considering cases of explanatory loss of the sort that interests Quincy and me, we are imagining situations in which a newer theory simply fails to address certain types of phenomena addressed by a previous theory. Hence it is not a situation in which T_2 fails by predicting a wrong result; it simply fails to say anything at all. That is the paradigmatic case of explanatory loss.

Realist: Thanks for the clarification.

Pragmatist: Now, as I was saying, in these circumstances would we have any grounds for holding that T_2 was more reliable than T_1 and thus that T_2 represented progress over T_1?

Positivist: I can see where you're headed, Percy. You're going to claim that, by virtue of the fact that T_2 has passed more demanding tests than T_1, T_2 represents progress over T_1—even if T_1 manages to have passed certain tests that T_2 was not subjected to.

Pragmatist: Precisely. One theory need not have passed all the tests of a rival for us to judge that it is better tested than, and thus represents progress over, its rival. If a later theory passes more robust tests than its predecessors, then we have good grounds for believing that the later theory will be more reliable than its predecessor, *even if* the earlier theory passed some tests which its successor did not.

Relativist: But what grounds have you for thinking that typical cases of explanatory loss in the history of science are always of this sort? Surely it sometimes happens that a later theory passes tests which are no more robust than those passed by its predecessor.

Pragmatist: To be honest, Quincy, I'm not sure how frequently either case occurs. The point I am trying to make, however, doesn't depend on relative frequencies of the sort you have in mind. What I am trying to show is simply this: the fact that earlier theories sometimes explain some phenomena not explained by their successors—and I do take that to be a fact—is by itself insufficient grounds for us to make the claim that science fails to progress in such episodes. If you want to deny that scientific progress occurs—and I think you do—then you must show (a) that the losses in question concern phenomena which were genuine tests of the earlier theories and (b) that the earlier tests were at least as robust as the tests passed by the later theories. I do not believe that you and your fellow relativists have undertaken either task.

Relativist: I grant you as much, but you are still missing the central philosophical moral of noncumulativity. The fact is that earlier theories generally solve problems not solved by their successors. Since, to paraphrase you pragmatists, the aim of science is to solve problems, and if two rival theories solve different problems, then it is a subjective matter to decide which theory is best, depending on our preferences as to which problems are more important to solve.[9] Progress in such matters is entirely in the eye of the beholder.

Pragmatist: Ever since Kuhn's *Structure of Scientific Revolutions,* you relativists having been making hay out of the fact that earlier theories sometimes solve problems not solved by their

9. Kuhn: "Since no paradigm ever solves all the problems it defines and since no two paradigms leave all the same problems unsolved, paradigm debates always involve the question: which problem is it more significant to have solved?" (1970, p. 110).

successors. But the point I have been trying to make for the last few minutes is that we do not judge theories primarily in terms of whether they solve some problems which we would like to have solutions for, problems to which we may even have assigned a high initial importance.

Realist: I find it a little rich, Percy, that you, as a self-avowed pragmatist, are trying to persuade us that solving important problems is unimportant in the appraisal of theories.

Pragmatist: I'm saying nothing of the sort. We develop theories because we find ourselves in problematic situations, where certain questions about the natural world forcibly impress themselves on us. Science, in my view, is entirely a problem-solving activity. But as an epistemologist, I am perfectly capable of distinguishing between those problems which arise out of salient practical concerns and those solved problems which constitute especially probative tests of a theory in question. Obviously, we want a theory to solve certain problems, but it would be a very naive pragmatist indeed who ignored the fact that the ability of a theory to solve the very problems it was devised to solve is usually not a very strenuous *test* of the theory in question, nor a very good indicator of how reliable the theory is, or how likely it is to hold up to further extensions and applications of it. The pragmatist is chiefly interested in whether a theory will be a reliable guide to the future. For that reason, he's apt to attach greater weight to theories that have passed robust tests than to theories which, while perhaps solving many problems, have passed few genuine tests.

Relativist: But if a theory fails to solve some problems we regard as important, we will surely reject the theory, regardless of how well-tested it appears to be.

Pragmatist: Wrong! If a theory fails to provide a solution to problems which we deem especially urgent or compelling, then we will, of course, cast around for some theory which does solve those problems. You are right to that extent. But the failure of a well-tested theory to solve problems that we want solved is no reason to reject the theory in question. If a theory is the best-tested among its rivals, that is, among the known contraries to it, then that must be the theory of choice among the rivals. Of course, if one can develop an equally well-tested, or better tested, theory which also solves one's preferred problems, all the better. But faced with a Hobson's choice between an ill-tested theory which solves a preferred problem and a well-tested theory which does not, the theory of evidence makes it very clear how one's choice should go.

Relativist: That is patently absurd. Are you trying to tell me that if, say, a chemist is interested in developing a theory about how colloidal suspensions work, and that if the best-tested theory of chemistry happens to be one which has nothing to say about colloidal suspensions, then he has no grounds for rejecting that theory, however well tested it is, because it fails to address what he regards as the central problems of chemistry?

Pragmatist: What I am saying is that there are two quite distinct issues which you are running together. One issue concerns the problems to which a theory offers an answer or solution. The other deals with the evidence we have that a theory is well-founded, or is likely to offer adequate solutions to the problems it addresses. Sometimes, theories fail on both counts: they neither address interesting problems nor do they have a successful track record vis-à-vis the tests to which they have been subjected. Both of us, I suspect, can agree that such theories should be rejected.

Relativist: Well, Percy, you know that I have doubts about whether one can ever make an objective decision to reject *any* theory.

Realist: Please, Quincy, one issue at a time. We have already agreed that we will give you a hearing later on your views about the multiple ambiguities of testing. But surely, for purposes of today's discussion, you can see the point Percy is driving at?

Relativist: Proceed.

Pragmatist: Well, as I was saying, we can ask of any theory both whether it is well tested and if it addresses problems we should like to see resolved. I was suggesting that if a theory either fails on both scores or succeeds on both scores, then every one of us here would agree about what to do. The apparent problem cases arise when a theory fails on one count but succeeds on the other. And here there is an important asymmetry. If a theory has been well tested, and passes all those tests, then it is an acceptable theory—by which I mean simply a theory likely to stand up well to subsequent tests and applications. And that judgment holds *even if the theory fails to address some of our favorite problems*. By contrast, if a theory appears to solve some of our favorite problems but has thus far not yet passed any demanding tests, then the theory is unacceptable because we have no reason to believe that it will be a useful guide in our future interactions with nature.

Relativist: You're saying that we should accept a well-tested theory even if it fails to address any problems we regard as central? That we can never argue against a theory on the grounds that—although well tested—it fails to grapple with the central issues?

Pragmatist: The fact that a theory fails to address certain key problems that interest us may well constitute grounds for seeking to develop a new theory which explicitly deals with the problems of concern—and which we will subsequently proceed to test; but if we are confronted by two contrary theories, i.e., theories such that they cannot both be true, one of which is ill tested but addresses interesting problems and the other of which is well tested but fails to address those problems, then it is the latter—not the former—which deserves our allegiance.

Realist: I think that we are beginning to lose sight of the charge to our committee. We began by saying that we were going to explore in this morning's session what the relativist had to say about scientific progress. We've heard a lot about what pragmatists, positivists, and realists think about that issue but very little about the relativist picture of progress.

Relativist: I'm glad you said that, Karl, for I was about to speak up. I'm not quite sure what to make of this morning's discussion from a relativist point of view. We relativists have generally tended to focus our critique of scientific progress on the claim that all theories of cognitive progress presuppose the cumulative retention of content from one theory to another. Because we deny that such retention occurs, we have regarded progress as a utopian idea in search of instantiation. I am a bit thrown by the turn of the discussion during the last hour because Percy has been insisting that a notion of progress can be held compatibly with acknowledging certain sorts of losses during theory change. That is an interesting tack to explore, but I'm afraid that it leaves me unmoved.

Pragmatist: Because?

Relativist: Your analysis presupposes—as you would be the first to admit—accepting a distinction between the instances which test a theory and the phenomena which a theory explains or entails. I don't have any problems with figuring out what a theory entails or explains; but I have massive difficulties with your notion of what tests or supports a theory. To put it in an extreme form, my view is that what tests a theory is simply what scientists decide to allow to test a theory. I think that there is no nonsubjective distinction between which instances do and which do not test a theory. For that reason, I do not see that Percy's solution to the problem of scientific progress is any solution at all.

Positivist: I cannot speak for the rest of you, but I have the feeling that we are beginning to repeat ourselves. May I suggest that we break for lunch and reassemble early this afternoon?

Relativist: I concur with Rudy's suggestion but I want to add one important point before we close this discussion. Percy has gone to some pains to argue that progress is an *instrumental* matter, a question of advancing toward the realization of certain ends. We have said little thus far about those ends. Do they, for instance, vary from scientist to scientist, epoch to epoch or paradigm to paradigm? Or are they fixed and transcendent? We relativists generally think that aims must be relativized to particular agents and specific contexts. And if that's right, then progress evaporates before our eyes, for what is progress for me may be nonprogress for you. Indeed, nothing could be more relativist than that! In any case, I think this is an issue which requires more attention.

Pragmatist: Quincy, you've just added one more item to our agenda!

2

Theory-ladenness and Underdetermination

DAY 1, AFTERNOON

Relativist: Gentlemen, I would like to begin our session this afternoon, if I may, by voicing some general worries about the course of this morning's discussion. One of the presuppositions which all of you evidently bring to scientific epistemology is the idea that there is a straightforward sense in which theories can be "tested." Percy and Rudy in particular spoke repeatedly this morning about partitioning theories into those which had been well tested and those which had not. Indeed, Percy's proposed solution to the challenge of explanatory loss rested on the viability of just such a distinction. It will come as no surprise when I say that I have grave worries about the whole testing enterprise. In my view, tests are highly inconclusive affairs. Single hypotheses and theories are never tested individually but only as parts of much larger networks of belief. Moreover, broad theories or worldviews have the resources to protect themselves from the results of any battery of tests. Moreover, even if the testing process is less ambiguous than I fear it is, there is a larger issue about testing which none of you is facing up to. I refer to the fact that tests are designed according to certain rules or methods; those rules and methods are not given us from on high. They are themselves simply socially sanctioned conventions, with no probative force. Until we have explored issues such as these—for they are the fundamental questions that divide us—I cannot see what profit there will be in focusing on specific problems of theory change or theory evaluation.

Pragmatist: Since our central charge is to get clear about relativism, I am inclined to believe that Quincy is right to suggest that we should get these issues up front in our discussions. The problem is, Quincy, that you have specified a very broad range of topics for discussion.

Positivist: Hopelessly so. I suggest that we ask Quincy to identify one or two issues for us to focus on this afternoon, with an understanding that we will attend to the others later in our conversations.

Relativist: Fair enough, I would suggest that we temporarily table my worries about the *status* of methodological rules and principles, about their conventionality and lack of grounding, and focus specifically for now on the question as to whether these rules of scientific method do anything to delimit choice.

Realist: What precisely do you mean by that?

Relativist: What I mean is that the rules of scientific method—or the rules of evidence, if you prefer—radically underdetermine choice between rival theories. Empirical evidence, which most of you regard as so unproblematic, is itself theory-laden; still worse, the rules which tell us what bearing the evidence should have on our theory choices are themselves systematically ambiguous. Because all that is so, the neat line you were earlier attempting to draw between theories which have passed demanding tests and those which haven't can't be drawn.

Pragmatist: May we, Quincy, try to order our discussions for this afternoon by taking up in order the two theses you just propounded? I mean first the argument about the theory-ladenness of evidence and, second, the claim about the ambiguity of rules for theory testing.

Relativist: Let's do it.

Positivist: I don't see where you are going to go with the question of theory-ladenness, Quincy. I grant you that underdetermination is a serious issue, but each of us here—whether positivist, realist, or pragmatist—admits that observation (and thus "evidence") is theory-laden. Indeed, it was positivists like Duhem and Neurath and realists like Popper and Grover Maxwell who were among the vanguard articulating the arguments for that thesis. What special argumentative mileage do you relativists now think you can get out of this tired old cliché?

Relativist: I'm resourceful enough to take my arguments wherever I can get them. What I don't think the rest of you have realized, however, is the range of epistemic implications that flow from the recognition that observation—the classic empiricist's bedrock—is itself infected by theoretical assumptions.

Realist: So far, Quincy, you've been making a lot of hyperambitious claims. I think it's time you gave us an argument or two.

Relativist: Attend carefully, Karl, for it is you realists above all who have failed to come to terms with the epistemic implications of the loss of the once pristine observational base. When I say that all observations are theory-laden, I mean simply that there is nothing we can say about the world which does not go well beyond what we are "given" by our senses. Every act of cognizing involves applying language or concepts. Our language, like our conceptual structures, pigeonholes experience in various ways. The categories in terms of which we carve up the world and make it intelligible to ourselves are not given by the external world but arise from, I suppose, earlier linguistic practices, our technical and practical interests as cognizers, and our built-in neurogenetic equipment.

Positivist: This sounds like warmed-over Kantian idealism to me, with a bit of evolutionary theory thrown in for good measure. The active role of the mind and all that. I repeat: we've heard it all before.

Relativist: The question is whether you've yet managed to come to terms with it, and if so how?

Positivist: I've no serious quibbles with what I've heard so far but it does strike me as rather vague.

Relativist: Then let me try to make it both more concrete and more threatening. If we once grant that every piece of "evidence" we can possibly have—every "observation report" as Rudy and his cohorts were once prone to call them—involves certain theoretical assumptions, assumptions not given by experience itself, then it should be clear that there is no sure empirical bedrock for our knowledge. Just as a priorism came unstuck because there is no sure starting point from above, then surely empiricism must be undermined by the recognition that knowledge is precarious all the way down—including those things which we purport to observe and measure.

Realist: I'm not sure whether empiricism in general is threatened by the recognition of the theoretical character of observation, but I wholly share Quincy's view that the constructive role of the mind in all forms of cognition gives the lie to narrow forms of empiricism such as positivism and instrumentalism. In fact, I have myself argued that those two theories of knowledge presupposed that theory could be sharply distinguished from observation and that observational claims were secure whereas theoretical ones were dubious. Indeed, the instrumentalist slogan, "save the phenomena," makes no sense unless the phenomenal or the observational can be clearly marked off from the theoretical.

Positivist: Some of the early positivists were, of course, too hasty in suggesting a sharp and fixed distinction between the observational and the theoretical; there is surely a continuum between them. But that needn't stand in the way of our recognizing the epistemic differences between respective ends of the spectrum. The claim that I'm now holding a pipe in my hand is surely on a different footing, in terms of our confidence about it, than, say, the claim that primitive organisms were first brought to earth on the surface of comets.

Relativist: The problem, Rudy, is that the sort of "evidence" used to support our fundamental physical theories—the sort of thing you refer to when you say that our theories are well tested—is rather more like the comet claim than the pipe claim. Just reflect on the amount of fine-tuning of theory and grooming of data that goes into producing the "evidence" which ostensibly confirms quarks or the general theory of relativity or plate tectonics.

Positivist: You're probably right about that.

Realist: Because there's no sharp distinction between theory and observation, the old myth that only what comes directly from observation is trustworthy has to be junked. And that fact has breathed new life into scientific realism in the last two decades; for it ushers in the recognition that theories are absolutely central in our conceptual scheme of things.

Pragmatist: I think, Karl, that you and your fellow realists may find that the sword cuts both ways.

Realist: What do you mean by that?

Pragmatist: It is you realists after all who argue for the primacy of theory; it is you who claim that we use theories to "correct" our observational claims; it is you who insist that we should be replacing commonsense or observational talk with deep-structure talk about the unseen causes of the world.

Realist: Yes, and so?

Pragmatist: The point is that nothing I just said is coherent unless you can draw a pretty clear distinction between the theoretical and the observational. If indeed *all* our claims about the world are theory-laden through and through, then it makes no sense to talk about how "theory" corrects "observation" or about how "theories" which explain more "observations" should be preferred to those which explain fewer. Indeed, you realists cannot even draw your crucial distinction between "deep-structure" claims and "surface" claims unless there is a pretty clear line between theory and observation. For such reasons, it seems to me that realist semantics and epistemology both continue to rest on the assumption that observations can be cleanly distinguished from theory; yet it is precisely the demise of that distinction which you think shores up realism and discredits instrumentalism.

Relativist: The challenge to realism posed by the theory-ladenness of observation is even more acute than Percy lets on. If our evidence reports do in fact rest—as you all grant they do—on the theories to which we now subscribe, and if you also concede, as history shows, that those theories are likely to be false in some respect or other—does it not follow that what we regard as our best evidence is almost certainly false in some respect or other? And if that is so, we are in a position to make a much stronger critique of the senses than the skeptic ever mounted. The most he could lay claim to show was that sensory reports about the world *might* be mistaken. Well, you silence his worries by admitting the point in principle but largely ignoring it in practice. That in effect is what fallibilism or corrigibilism comes to. What I am now telling you is that what we regard as evidence is not just possibly mistaken, it is almost certainly wrong; and this is because the theories on which the evidence rests are almost certainly wrong. And if that is so, then the idea of testing our theories by submitting them to the "impartial" tribunal of evidence is a will-of-the-wisp.

Pragmatist: You're suggesting that testing a theory does no more than bring one set of theories—those ostensibly under test—to bear against another set of theories—those utilized to generate the evidence in the first place?

Relativist: Precisely. And the subsequent judgment that the "'evidence' *supports* the 'theory'" just reflects the logical compatibility between the theories under test and the theories undergirding the evidence. Equally, the judgment that the "'evidence' *refutes* the 'theory'" is simply the determination of an incompatibility between two sets of theories. Testing is therefore no confrontation between our theories and the world. It is rather an exploration of the mutual compatibility of two sets of theories.

Positivist: Let me be clearer about your argument. You are claiming, 1, that all our theories are eventually likely to be superseded by other, contrary theories, hence the former should be presumed to be false, even if we don't know it yet; 2, that every report of an experiment or observation logically presupposes some subset or other of current theory; 3, that the likely falsity of our theories constitutes grounds for inferring the likely falsity of our observation reports; and, therefore, 4, that the use of "theoretically contaminated" (and thus presumably false) evidence to "test" our theories is of no probative significance—except insofar as it establishes whether the totality of our current theories is internally consistent. Have I got it right?

Relativist: Exactly.

Pragmatist: I'm curious, Quincy, about the status of what Rudy has just called premise 1. I agree with the claim that most current theories are probably eventually going to be refuted; but for the life of me I don't see how *you* can make that claim, consistent with your other beliefs. Your argument for that claim, if I remember what you said just five minutes ago, was that we have only to look at the history of science to see that theories eventually get falsified. Am I right that it is the repeated record of such falsifications that provides your license for claiming premise 1?

Relativist: Yes; it's what is called the "historical induction" or sometimes the "pessimistic induction."[1]

Pragmatist: But while you're telling us that we have this massive historical record of theory falsifications, you are also asking us to believe that evidence is invariably suspect and ambiguous. You are also the one who tells us that theories can never be refuted.

Relativist: Indeed, although I haven't yet had a chance to set out the arguments for nonfalsification.

Pragmatist: We'll get to that soon enough. But for now my question is this: How are you entitled to assert *both* that theories are irrefutable *and* that we should deem current theories false because past ones have been shown to be?

Relativist: I am making this argument about the epistemic morals of the theory-ladenness of observation not, as it were, within the frame of my own presuppositions, but within those pervasive among mainstream epistemologists and philosophers of science. What I am saying is that all of you believe what Rudy called my first premise, to the effect that all current theories are probably false. That's right isn't it? Hearing no denials, I'll go on to suggest that each of you also

1. Putnam (1978) and Laudan (1984) use these terms respectively.

believes premise 2—that all evidence is theory-laden. No dissenters? Good. As for steps 3 and 4, they seem self-evident if we grant 1 and 2. What I purport to show by this general argument is that those beliefs commit you to regarding tests as nonprobative. Of course, I share that conclusion, although I come at it myself from a different direction.

Realist: Since we all do accept your premises 1 and 2, the real issue is whether we are thereby committed to 3 and 4.

Positivist: I for one am very perplexed by 3, the claim that observation reports are likely to be false because the theories they presuppose are probably false. That strikes me as patent nonsense. Suppose that I am testing some theory about the structure of star clusters and that I use optical telescopes to collect the relevant evidence. Now obviously, how I interpret and understand my observations is going to depend in part upon the optical theories I have about how light is transmitted through a telescope. Let us suppose, for the sake of argument, that those optical theories eventually turn out to be false. My question is this: does the falsity of those optical theories necessarily invalidate the observations made under their sponsorship, as it were?

Relativist: Of course it does. Very different understandings about what is actually happening to light inside a telescope can dramatically effect one's construal of what one "sees" by means of the telescope.

Positivist: But surely that depends on *the respects in which the optical theory turns out to be false or otherwise unacceptable*. It is surely conceivable that our optical theory may eventually break down with respect to some highly esoteric optical phenomena (say certain forms of polarization of light), a breakdown that may do nothing to impugn those bits of theory that were at stake in interpreting the movement of light through our telescope. Indeed, the lower-level "laws" associated with our theories often manage to survive the demise of the deep-structure theories with which they were once associated.

Relativist: Not intact they don't. They are typically modified, sometimes in subtle ways.

Pragmatist: Come on, Quincy, the law of reflection has remained virtually unchanged since antiquity; and most of the dioptric and catoptric laws developed in the seventeenth century remain the ones in practical use in our time—despite our rejection of the theories with which those phenomena were first associated.

Relativist: I thought we had fought out this particular battle this morning, Percy, and you were on the other side then. You recall that we established the fact that because different theorists subscribe to the same or homologous equations, this does not mean that they construe the terms of their theories in similar ways. Just consider the differences between a wave theorist and a particle theorist, who will understand the laws you were just describing in *fundamentally different* ways.

Pragmatist: Granted. But the issue at stake here is not one about what deep-structure stories the advocates of different theories will tell about the laws they share in common. What we are discussing is your premise 3, to the effect that the likely falsity of our theories creates a presumption in favor of the likely falsity of the observation reports which draw on those theories. I think that Rudy's point to you was that, when a once well-established theory breaks down, it is typically not in a wholesale fashion but because one element or other of it fails. If that picture of theory demise is right, and I for one find it congenial, then we ought not so readily slide from the presumed falsity of a theory to a presumption of the falsity of all of its constituent parts.

Positivist: I think we can make the point pack more punch than that. Let me remind you, Quincy, of one of the favorite arguments of you relativists and show you how that argument undercuts your view on the theory/observation business.

Relativist: And what is that?

Positivist: Well, the so-called Duhem-Quine thesis involves the argument that one cannot deduce the falsity of any one element of a network of statements from the falsity of the network as a whole.

Relativist: And so?

Positivist: Well, in arguing that observations which are linked up with a particular theory should be abandoned when the theory is abandoned, you seem to be supposing that the falsity of a theory as a whole infects each of its constituent parts.

Relativist: I don't see that I'm committed to that.

Positivist: Let me put it this way. The Duhem-Quine thesis says that hypotheses never confront experience singly but only as parts of larger packages, involving other hypotheses, initial and boundary conditions, and the like. Right?

Relativist: Yes, go on.

Positivist: According to Quine, what gets refuted is the entire complex of assumptions used in generating a mistaken prediction. Quine and other relativists are adamant that we cannot draw any inferences from the failure of the complex as a whole about the falsity of any one of its constituents. Agreed?

Relativist: Of course, but I wish you would get to the point.

Positivist: The point is that, by your own lights, we are not entitled to draw any conclusion about the falsity of a particular hypothesis from the falsity of a larger complex of which that hypothesis is a part. Yet you are proposing we should do just that. You would have us believe that since, say, the wave theory of light has been refuted, we are to suppose the falsity of each of its constituent elements.

Relativist: I said no such thing.

Pragmatist: Ah, but you did. For you told us that since most global theories in the history of science have been refuted, and since virtually every observation depends on assuming some theoretical hypothesis or other, we had reason to expect that all our observation reports were also false. That argument makes sense *only* if we suppose that the refutation of a large theory involves the discrediting of *each* of its constituent parts; and the latter result is at odds with the Duhem-Quine thesis. Rudy and I have been trying to show you—if you are the fan of the Duhem-Quine thesis that you claim to be—that you should readily grant our point that some of the "parts" of a refuted theory are unimpugned by the demise of that theory in general, and that the observations which depend upon those unchallenged components may remain reliable.

Relativist: That's all very well, Percy, but you and Rudy continue to ignore my point that once *any* part of a theory is changed, there are subtle changes in meaning in *all* the terms constituting that theory. And because that is so, any change in a theory is going to have ripple effects which spread throughout the whole theory and its associated laws. That in turn means that current views about the laws of nature and about the meaning of specific "observations" are as likely to be abandoned as are theories. All of which suggests that the meaning of any putative evidential claim is hostage to the specific cluster of theories with which it is currently associated. Change those theories—as we are bound to over the course of time—and you thereby change the sense of the evidence.

Positivist: You're asking us to reject as false Galileo's observations about Jupiter having moons *because* we reject as false Galileo's theories about how telescopes work. That is patently absurd.

Realist: What it appears to come down to, Quincy, is whether we are willing to accept a certain theory of yours about how terms acquire meaning.

Relativist: I suppose that's a perspicacious way of putting it. I see meaning changes flowing throughout a theory's network of assumptions as modifications are made anywhere in the system; whereas you apparently hang on to some as-yet-unspecified account of the meaning of scientific terms which purports to block the trickle-down effects of those shifts. I wonder, Mister Chairman, if this isn't an issue to which we should devote one of our future sessions?

Pragmatist: I've just jotted down that we should do so.

Relativist: In those circumstances, I am prepared for the sake of today's argument to proceed on the assumption that specific evidential claims are not automatically contaminated when the theories with which they are associated are rejected.

Positivist: Hooray, progress at last!

Relativist: But there is another side to this issue which I think you are all ignoring. When Percy asked me earlier how I, as a card-carrying relativist, could happily talk as if theories were refuted by the evidence, I responded by saying that I was speaking your vernacular rather than mine. Do you remember?

Realist: Yes, I do.

Relativist: Let me now formulate my worries on this score in my terms rather than yours, since I think your apparent victory here may be pyrrhic once you realize what's actually at stake.

Pragmatist: Do continue, but be as succinct as you can.

Relativist: When I use the phrase "theory-laden observations," I mean to focus on more than meets the eye. . . .

Positivist: That was a terrible pun.

Relativist: Unintended, I assure you. What I mean is that there is an element of *convention* in every observation statement, and doubly so. One sort of convention associated with observation statements is simply the decision to accept a given "observation statement" as correct. There is nothing in our experience of the world which demonstrates infallibly that the world is as we perceive or observe it to be. All of us know that purported observations are often found to be faulty, in need of revision, etc. Both Quine and Popper, among others, have stressed that the decision to accept an observation statement as a veridical report of how the world is involves a defeasible decision.

Realist: I'll buy that, but telling arguments can be given as to why, if one is an empiricist, one should suppose that reports of observation are to be given preference—other things being equal—over statements of theory.

Pragmatist: Not wanting to divert the course of the discussion, I cannot resist reminding our realist friend that he is not supposed to believe in the distinction between the theoretical and the observable.

Realist: I think I'll let that one pass.

Positivist: Please do! Quincy, you were earlier telling us that there are two sorts of conventions associated with observations; one, as I understand it, is the decision to accept them as foundational. What is the other?

Relativist: The other has to do with language itself. Whenever we report an "observation," we couch it in some language or other; and all of us know that languages pigeonhole the world in ways that are highly conventional. If I say, for instance, that this desk is 2.2 meters long, I am supposing that objects like desks are genuine kinds of things, that meters reflect some natural unit for measuring a real property called "length," and so on.

Pragmatist: And what follows from all this?

Relativist: Well, take my example and reflect on it for a minute. We all agree that the length of a meter was settled by convention. Right?

Pragmatist: Of course.

Relativist: Hence any statement reporting length in terms of meters will be inherently conventional. There is no fact-of-the-matter about the world which says that it must be measured in meter-type units.

Positivist: That's surely correct. Science, or any other form of inquiry, begins with certain definitions such as the definition of a meter and of the operation constituting "length measurement." But I'm with Percy on this one: so what?

Relativist: Well, a meter is a convention—one that reflects no facts-of-the-matter; it follows that any statement involving reference to meters will be similarly conventional. And if conventional, it is neither true nor false. That's what I meant earlier when I said that falsifications are arbitrary. If observations are neither true nor false, they can scarcely falsify a theory or anything else.

Realist: You're going much too fast for me, Quincy. I agree with you that the definition of a meter-length depends on convention. But when I say that this desk is 2.2 meters long, I am making a claim about the facts-of-the-matter. I am claiming that the length of this desk is a little more than double the length of the standard meter stick in Paris. That claim is either true or false; it is not a convention, even though it is formulated in terms of conventions.

Relativist: You're saying that although these notions—meter, length, and desk—are all conventions through and through, we can make claims using these concepts which are not conventional?

Realist: That's precisely what I'm saying. It is possible, using a system of conventions, to make claims about the world which are not matters of convention.

Relativist: But the conventions could have been otherwise—that's what makes them conventions!

Realist: Indeed they could. But even if we changed the meaning of meter to mean (say) the length of William the Conqueror's arm, it would still be either true or false to assert of various things in the world that they were various multiples or fractions of that length.

Pragmatist: I'm not sure that we're making a lot of headway on this one.

Relativist: I quite agree. If I may, I'd like to go back to something that Karl said a few moments ago, for I think it puts some of our differences quite succinctly. He was talking about why, when there is a conflict between our theories and the evidence—a conflict which we all seem to be willing to construe as a clash between two corrigible theories—the empiricist is committed to giving up the theories and accepting the evidence. I agree with him that empiricists are committed to just such a principle. But this appears to me, in light of our preceding discussion, simply to be one more of the dogmas of empiricism. None of you has given any arguments for this priority.

Positivist: No we haven't; not yet at least. But I had supposed that was unnecessary in this company since I thought that even you, Quincy, were a committed empiricist. It is you relativists who claim to use evidence from the history of science to support your theories and to discredit ours; it is you who claim that epistemology should turn itself into a descriptive social science. None of that makes any sense unless you yourselves believe that evidence takes priority over theory when there is a conflict between the two.

Pragmatist: But Quincy's challenge remains on the table, unanswered except by an ad hominem argument. I think it deserves a reply.

Positivist: I'll be happy to give one, and I promise to be brief. Let us remind ourselves of the situation we find ourselves in. We're agreed that theories are involved in the construction and interpretation of instruments; we agree that theoretical assumptions go into determining the boundary conditions which we suppose to apply to any situation under scrutiny. Let's call all that baggage, for ease of reference, the *observational* theories. These are to be contrasted with the theories which are being scrutinized; let's call them the *target* theories. When what we used to call an empirical refutation occurs, we could now describe it as a clash between an observational theory and a target theory. Quincy's challenge to the empiricist, as I see it, is to explain why—in such circumstances—we should sacrifice the target theories rather than the observational theories. Have I got that right?

Relativist: Quite.

Positivist: In that case, the answer is simple. We should do here what we do *whenever* we are confronted by a choice between two mutually inconsistent theories. We ask: is one of these theories much better supported than the other? If we can give an affirmative answer to that question, then we know which to reject. In the case of clashes between observation theories and target theories, the observational theories are generally much more solid than the target theories. That is why, confronted with a clash between an "observation" and a "theory," we generally reject the "theory."

Realist: That is much too quick, Rudy. As Sellars and Feyerabend have argued, we sometimes use our more speculative theories—what you are here calling target theories—to correct our observational hypotheses.

Positivist: So we do; nor would I wish to be taken as denying that. There are occasions when target theories are better established—by which I mean better tested—than the observational theories. But I daresay that this is the exception rather than the rule. Shrewd scientists attempt to design their experiments in such a way that the observational theories required are more dependable than the target theories under test. Failure to do that undermines the claim that a test has occurred. It seems to me that this provides a perfectly general solution to the problems Quincy thinks are generated by the theory-ladenness of observations.

Relativist: So here we are, again back to square one. As in this morning's discussion, you three, by invoking a distinction between what has been well tested and what has not, are attempting to avoid the paradoxes to which your positions lead. I think we cannot afford to postpone any longer a discussion of my claims about underdetermination; for it is those which constitute a direct challenge to this notion of being well or robustly tested. Specifically, I can show that, even if observation statements are not conventional (in my second sense), and even if observations fail to be casualties of the false theories with which they are associated, I can still show that "test" results ***radically underdetermine*** the choice between rival theories. The arguments on this issue are the linchpin on which much contemporary relativism depends—including the theory of meaning to which I was alluding earlier this afternoon. I think that we should not postpone a consideration of the core case for relativism, since so many of the points I am tempted to make in these exchanges require a familiarity with issues of underdetermination.

Pragmatist: Since I wholeheartedly agree with your claim that the thesis of underdetermination is absolutely central to epistemic relativism, I would be delighted to turn to it. Anyone quibble with that? . . . In that case, Quincy, begin the development of your case.

Relativist: The argument is really very simple, as all genuinely deep arguments are. It begins with the claim that we are never in a position to have access to more than a finite number of observation reports, measurements, etc. It then proceeds to show that there are always indefinitely many contrary hypotheses or theories compatible with any finite set of observations.

Positivist: In other words, we cannot deduce theories from observations?

Relativist: Yes, you could put it that way.

Positivist: Well, there was malice in my formulation since the nondeducibility of theories from finite collections of their instances has been a truism at least since the time of Hume, if not long before.[2] I daresay that all of us have managed one way or another to come to terms with the fact that theories are not deducible from the phenomena. What new light have you to shed on this issue?

Relativist: It is the implications of this thesis that most of you, especially realists like Karl, have yet to deal with. For what the thesis of underdetermination tells us is that, no matter how extensive our acquaintance with the natural world, there will remain indefinitely many—arguably infinitely many—incompatible theories all of which are equally compatible with the available evidence. Indeed, even if we had a God's eye perspective and could ascertain that all of a theory's potential observational consequences were correct, we still could not tell if the theory were true since it would have indefinitely many rivals with the same observational consequences. . . .

Positivist: Stop right there, Quincy, if you don't mind. I know that you are developing a criticism of realism and doubtless Karl will soon want to put his oar in; but I'm troubled by something else. I was jotting down some notes as you were sketching out your position and I thought you said—correct me if I'm wrong—that the thesis of underdetermination amounts to the doctrine that there are indefinitely many contrary theories, all "compatible" with any finite set of observations.

Relativist: Yes, that's exactly what I said.

Positivist: But you were just now telling us that even if we had a God's-eye perspective, we would still find that indefinitely

2. For an argument for the pre-Hume ancestry of this problem, see Laudan (1981).

many different theories would have the same observational consequences.

Relativist: And so?

Positivist: Well, which is it to be: does the thesis of underdetermination assert that indefinitely many theories are "compatible" with any finite body of evidence *or* that indefinitely many theories "entail" that evidence?

Relativist: Ah . . . , yes, I see your worry now. There is, of course, a difference between being compatible with the evidence and entailing the evidence. I think that the underdetermination thesis surely asserts both.

Realist: You'd better believe there is a difference! The claim that indefinitely many different theories are compatible with the evidence is an extraordinarily weak doctrine. It's doubtless true but of decidedly limited epistemological interest.

Relativist: Why do you say that?

Realist: Well, all the brouhaha about underdetermination arises because people like you suppose that saying theories are underdetermined amounts to saying that we can never give compelling reasons for the choice of one theory over another. Am I right?

Relativist: Of course.

Realist: As I see it, the fact that two theories are *compatible* with all the available evidence is *no* reason for us to think that we cannot rationally choose between them. Suppose that one of those theories asserts that "God is a trinity" while the other is the kinetic theory of gases. Suppose further that our evidence is drawn from observations on the behavior of gases. Under those circumstances, the trinitarian hypothesis will presumably be compatible with everything we have observed about gases; but that bare compatibility will give us no grounds for asserting that the evidence gives reasons for accepting the trinity hypothesis that are as compelling as those it gives for accepting the hypotheses of statistical mechanics.

Positivist: Karl's point, Quincy, is that the *compatibility* of a hypothesis with the available evidence offers no positive grounds for holding that the evidence thereby *supports* the hypothesis with which it is compatible. If the underdetermination thesis is to have any punch against the realist or anyone else, you have to make it stronger than the claim that indefinitely many theories are compatible with any finite body of evidence.

Relativist: I can see that clearly enough. But as I said a few minutes ago, the underdetermination thesis similarly asserts that indefinitely many theories can explain the same body of evidence.

Pragmatist: Did you say "explain" or "entail"?

Relativist: I actually said "explain" but it comes to the same thing in these circumstances. Indefinitely many rival theories entail and thereby explain any finite body of data.

Pragmatist: Are you trying to tell us that explaining and entailing amount to the same thing?

Realist: Well, to join the issue on Quincy's behalf, I should remind you that the classical deductive-nomological model of explanation makes precisely that assumption, to wit, that if *x* explains *y*, then *x* entails *y*. I realize that there are various statistical forms of explanation that do not involve entailment, but I think we can ignore such complications in this discussion.

Pragmatist: So we can, but the problems of statistical explanations are not what I was driving at. You're quite right that Hempel et al. argued that the archetypal form of explanation was one in which the premises both entailed and explained the conclusion. What I was challenging was Quincy's supposition that all cases of entailment were ipso facto instances of explanation.

Relativist: I don't seem to remember putting the point that way.

Pragmatist: What you actually told us was that, according to the thesis of underdetermination, indefinitely many theories would entail any given body of evidence, from which you inferred that indefinitely many theories explained that evidence. To argue in that manner is to suppose that a theory explains all the statements it entails.

Relativist: That seems fair enough.

Pragmatist: You'd better think again, Quincy. For unless we hold that theories do *not* explain everything they entail, we end up in some messy paradoxes. The most obvious one is self-explanation.[3] Clearly, a theory entails itself yet few of us would be so bold as to claim that a theory explains itself. Such intuitions show that there is a difference between the set of a theory's entailments and the set of its explananda.

Relativist: I grant you that point, but it strikes me as fairly arcane philosophical nit-picking.

Pragmatist: Quite to the contrary, very large issues are at stake here. You have been claiming, a, that indefinitely many theories have the same observational consequences. In due course, I shall remind you that you have not yet proven that assertion but let's give it to you for now for the sake of the argument. You went on to say that this showed, b, that indefinitely many theories can explain the same phenomena. Now, if you will grant me (as you just have) that entailing a statement, *s,* and explaining *s* are two different things, then I have to emphasize that from the supposed truth of a, you cannot derive b.

Relativist: The fact is, Percy, that I don't need a result as strong as b. It is sufficient for my purposes if I can show a, for if a is correct, that means that we are never going to be in a position to certify a theory as true, even if all its examined logical consequences square perfectly with what we observe. As I was trying to say before this long interruption, that fact should give serious pause to realists who tell us that the aim of science is to find true theories.

Realist: Why should the fact, if fact it be, that we can never certify a theory as true stand in the way of my claim that we are seeking true theories?

Pragmatist: Since Quincy has been getting more than his share of grief lately, let *me* try to answer that if I may. If we propose an aim for an activity which we know we will never be in a position to certify as satisfied, or as getting closer to satisfac-

3. For a discussion of many of the drawbacks to the assimilation of explanation to deduction, see A. Grünbaum et al. (1988).

tion, then we have no way of telling whether we are making progress in our endeavors to realize that end. To tell scientists that they should seek true theories when you concede that there is no way of certifying any theory as true, or "truer than another" is to enjoin them to engage in a quixotic enterprise.

Relativist: But it is not only the realist who should be embarrassed by the implications of underdetermination. So too should you pragmatists and positivists. Pragmatists characteristically hold that the "true" is the maximally useful. But if "useful" means something like being effective at saving the phenomena, then the pragmatist is forced by virtue of the thesis of underdetermination to hold that indefinitely many contrary claims can simultaneously be "true." And that is hardly a happy position to be in. Similarly, the positivist should feel discomfited by the thesis of underdetermination since that thesis. . . .

Realist: Before you continue with this dreary catalogue of debilitating consequences that flow from the thesis of underdetermination, I wonder if we oughtn't to examine the thesis itself? So far, Quincy, you have stated the thesis but you've given us no arguments in its favor. I wonder if we couldn't go through those arguments a step at a time.

Relativist: I'll gladly rehearse that terrain if you like, but I do seem to remember Rudy telling me impatiently that the arguments for underdetermination had been familiar ever since Hume, so I didn't want to bore you by taking you through them one more time!

Realist: Well, I can't speak for my colleagues, but why don't you humor me by setting out, as you see it, precisely what underdetermination is and why it carries the epistemic freight which you think it does.

Relativist: My pleasure. Indeed, I'd like to take the opportunity to state the thesis of underdetermination in a logically stronger version than hitherto.

Positivist: But you haven't yet persuaded us to accept it even in its current form!

Relativist: All in good time, Rudy. Up to now, I've limited the claim of underdetermination to the assertion that, for any theory that is, in your terms, well tested, there will be indefinitely many rivals which are equally well tested. I want now to expand that claim by asserting that *all* the rivals to any given theory are as well supported by the evidence as the theory in question is. From that it follows that we have no epistemic grounds for accepting or rejecting any theory rather than any of its contraries.

Positivist: You're claiming then that *all* rival hypotheses we might imagine are on the same evidential footing? That position is patently absurd. As one of your fellow relativists—Richard Rorty—points out, nobody really believes that.

Relativist: Rorty notwithstanding, I think that most contemporary relativists—and many folks who don't even admit to being relativists—believe precisely that. If you have doubts about it, consider the fact that Kuhn has claimed repeatedly that there is no point, in the accumulation of evidence and argument, at which it becomes "unscientific" to continue to hold onto a paradigm.[4] If he really means that, I suppose that he intends to be saying that the advocacy of Aristotelian physics, Galenic medicine, phlogistic chemistry, etc., would be just as "scientific" as advocating their more popular contemporary counterparts.

Pragmatist: And I seem to remember hearing it argued that Quine is committed to something very similar. After all, it was he who said, in his classic "Two Dogmas of Empiricism," that "*any* statement [or theory] can be held true *come what may*."[5]

Relativist: Or recall Lakatos's quip that "A brilliant school of scholars (backed by a rich society to finance a few well-planned tests) might succeed in pushing *any* fantastic programme [however 'absurd'] ahead, or, alternatively, if so

4. Apropos of resistance to the introduction of a new paradigm, Kuhn claims that the historian "will not find a point at which resistance becomes illogical or unscientific" (1970, p. 159).

5. Quine specifically put it this way: "Any statement can be held true come what may, if we make drastic enough adjustments elsewhere in the system [of belief]" (in S. Harding, 1976, p. 6; my emphasis in text).

inclined, in overthrowing *any* arbitrarily chosen pillar of 'established wisdom'."[6] Now, if these statements mean what they say, they are asserting that any theory whatever is as good as any other.

Realist: I think Feyerabend and some sociologists have said something along the same lines.[7]

Pragmatist: In any event, Quincy, the issue surely is not how many people have made such a claim but whether you have any good arguments for it.

Positivist: I'm as eager as you are to hear Quincy's arguments, Percy, but there's one preliminary I still would like to have cleared up. For me, it is enough for the relativist to argue that many theories will be on a par evidentially. I can't for the life of me see why any relativist would want to take on the more ambitious claim that virtually all the rivals to any given claim are equally well supported. I don't dispute that many scholars do put forward the stronger claim—as some of the passages we've just heard show. But I'm trying to understand why people would want to take on such an ambitious form of relativism.

Relativist: It's a fair question and I will try to answer it briefly. We might distinguish between weak and strong relativism. Weak relativism would be the position that there are some occasions when existing evidence does not permit a choice between certain rival theories. Strong relativism, by contrast,

6. In Lakatos and Musgrave, eds. (1970, pp. 187–88; my emphasis). Lakatos, despite his oft-proclaimed antirelativism, frequently lapses into language of which any thoroughgoing relativist could be proud. In one especially salient passage, he writes: "The direction of science is determined primarily by human creative imagination and not by the universe of facts which surrounds us. Creative imagination is likely to find corroborating novel evidence even for the most 'absurd' programme, if the search has sufficient drive" (1978, vol. 1, p. 99).

7. See also David Bloor's remark that "So [sic] the stability of a system of belief [including science] is the prerogative of its users" (1982, p. 306). Bloor's fellow conspirator, Barry Barnes, makes an identical claim: "Scientists, if they so desire, can always retain their verbal culture [by which Barnes means their theories inter alia] as an unfalsifiable system" (1982, p. 76), and: "A whole conceptual fabric can always be *made out* as in perfect accord with experience, if the scientific community sustaining it is of a mind to do so" (p. 75).

would hold that evidence is always powerless to choose between any pair of rivals. Rudy's question, if I've got it right, is: why be a strong relativist rather than a weak one? Right?

Positivist: Exactly.

Relativist: The answer is pretty simple. And it comes in two parts. The first part is that there is no one around these days who disputes the thesis of weak relativism. Positivists, pragmatists, even realists agree that there are some circumstances in which the available evidence fails to warrant a choice between certain rival perspectives. To propound such a thesis would evoke nothing but yawns. Second, and more important, the weak thesis of relativism requires me to make some concessions which would quite undermine relativist epistemology in general.

Realist: Such as what?

Relativist: Well, if I were once to grant you that one particular theory or hypothesis really had been shown to be better than a rival, I should thereby be conceding that there is some universal, timeless standard for the evaluation of beliefs. That's not a point I want to concede. I would also be conferring on evidence an unproblematic status which is completely at odds with my general perspective. There is more I could add here but I assure you that I have thought through this issue with some care and I'm persuaded that relativism stands or falls only on its stronger version.

Pragmatist: I don't think anyone will accuse you of taking up an easy brief. Perhaps we could now get to the solid arguments you promised us on behalf of full-blown relativism?

Relativist: The arguments are not only solid but they have the virtue of being straightforward. I begin by reminding you of Hume's claim to the effect that no genuinely universal statement (and I mean such statements when I talk about "theories" or "hypotheses") can be deduced from a finite set of its positive instances. Because that is so, no theory can ever be proved to be true.

Positivist: Big deal.

Relativist: Moreover, and this point was made by Duhem almost a century ago, we can no more derive the falsity of a theory or hypothesis from the evidence than we can derive its truth.

Realist: You're referring, I suppose, to the holistic character of the testing situation and the fact that we bring many assumptions to bear in the design of any test of a single theory or hypothesis.

Relativist: Exactly. It is never single hypotheses or theories we test but entire networks of such assumptions. When the test fails, that is, when the predicted result differs significantly from the observed result, all we know, at best, is that we have made a mistake somewhere; it cannot be further localized.

Pragmatist: I have serious doubts about this holism of yours, which I hope I will have the chance to voice later; but suppose we grant you, for now, both the Hume and the Duhem theses about the nonderivability, and the nonrefutability, of theories from their positive, or negative, instances. How does that establish your thesis that it is as reasonable to accept any theory as any other?

Relativist: I would have thought the connection was more or less self-evident. If we can never show that any theory is true and we can never show that any theory is false, then surely it is clear that whatever grounds we may have for choosing between rival theories must be practical rather than epistemic, or simply matters of convention and simplicity. Since there is nothing in the evidence which forces us to believe one thing rather than another, we can choose our beliefs in the light of our private and personal interests.

Realist: I have been biding my time while Quincy talks this gibberish because I wanted to give him a chance to have his say. But I really think it is time to call his bluff. His whole analysis rests on the assumption that we are entitled to accept or reject a theory only if it has been proven—or disproven. Hasn't he heard that we're all fallibilists these days and that no one is claiming that either confirmations or falsifications establish once and for all the truth or falsity of our theoretical beliefs? We don't have to be able to prove that a theory is true before we have a warrant for accepting it nor do we have to demonstrate its falsity before rationally rejecting it. All that the work of Duhem and Hume shows is that deductive logic has insufficient resources to allow us to do natural science. That would be a troubling concession for realists like me to make only if I thought that deductive logic exhausted our epistemic resources for talking about the relation between theory and experience. But we all know better than that.

Relativist: I'm not as out of touch with your views as you might think, Karl. I wanted to start where I did, namely, with deductive relations and their inability to motivate theory choice. And I am happy to talk with you in due course about the other, nondeductive resources of ours to which you just referred. Yet for the record I would like it clearly noted that all four of us are agreed that neither the truth nor the falsity of theories can be deduced from the evidence. Agreed?

Positivist: No, I don't agree. I fully accept that the truth of theories cannot be deduced from evidence and I am perhaps persuadable that neither can their falsity; but I haven't heard those arguments yet and until I do, I will be party to no such agreement.

Relativist: Very well, but you'll come around eventually, Rudy. In the meantime, I wonder if we can leave matters of deductive logic to one side and pick up on Karl's claim that there are other, nondeductive resources which are supposed to narrow theory choice. I conjecture that Karl has in mind something like the rules of inductive logic—which have themselves been discredited over and over again as a coherent system for assessing our beliefs. Carnap tried that tack decades ago and the whole thing came unstuck. More recently, the multiple paradoxes of inductive logic raise acute doubts about whether we can ever realistically hope to articulate a set of inductive rules, comparable to the rules of deductive logic, for assessing claims about the world.

Realist: I'm as skeptical about the prospects for a formal system of inductive logic as you are, and probably with better reason. It was, after all, scientific realists like Popper who exhibited the incoherence of the inductive logic program.

Positivist: And, to be clear about the historical record, it was positivists like Hempel and Carnap who first pointed out the paradoxes of induction.

Realist: Be that as it may, I do not see that my program for talking about levels of empirical support—short of proof and disproof—stands or falls with the viability of inductive logic per se. I think we have certain rules or yardsticks for evaluating theories. If a theory satisfies those rules better than its rivals, we have good reasons for accepting it; if it fails to satisfy those rules, then we have good reasons for rejecting it. Even though in neither case have we a proof or a refutation.

Relativist: And just what are these rules that you have in mind?

Realist: This is not the time or place to give you a full list, but it would include principles like the following:

—reject theories which fail to fit the known phenomena;
—prefer theories which make successful surprising predictions over those which fail to make surprising predictions;
—prefer theories which explain broad ranges of different kinds of phenomena to those which explain only phenomena of the same kind;
—if a theory emerges which offers the only explanation for certain phenomena, then accept it;

and so forth.

Relativist: But all these rules underdetermine theory choice in precisely the same way that the rules of deductive logic do. Take, for instance, your second principle about surprising predictions. Are you maintaining that the ability of a theory to make surprising predictions successfully is a reason for accepting the theory as true?

Realist: Well, there are some realists, like Popper, who hesitate ever to say that we are entitled to "accept" a theory. But most of us tend to side with Sellars and one of the many Putnams when they argue that a record of successful surprising predictions provides good, if corrigible, grounds for accepting a theory as true. So the short answer to your question is "yes."

Pragmatist: Surely, Karl, the history of science is full of theories which have successfully made surprising predictions—Newtonian mechanics, wave optics, classical electromagnetic theory—yet which we now regard as false. Your rule would have sanctioned accepting all of them as true.

Realist: I grant that point. But both you and Quincy are acting as if the ability to make surprising predictions successfully was for me a *sufficient* condition for accepting a theory. I regard it rather only as a necessary condition.

Relativist: But I don't see how that gets you off Percy's hook.

Realist: Quite simply. The claim—a correct one—that we have had false theories which made successful, surprising predictions would count against the predictivity rule only if I were saying that rule alone were sufficient to certify theories as true.

Positivist: But if your predictivity rule is only a necessary condition for an acceptable theory, how can we ever tell when we have an acceptable theory? In effect, you're telling us now that making successful predictions is not enough to justify us in accepting a theory. Something more is surely required.

Realist: And I've already hinted about what that is. You'll remember that I suggested several rules which theories must satisfy before we're entitled to accept them as true.

Relativist: Your view then, if I've got it right, is that there are various constraints, call them *a, b, . . . , n,* which are individually necessary and jointly sufficient for certifying a theory as true?

Realist: Exactly. Provided, of course, you understand that acceptance is defeasible.

Relativist: Let's call that set of requirements by the term "robustness," and say that on your view a theory which exhibits robustness is acceptable as true. Okay?

Realist: Fine.

Relativist: But Karl, don't you see that all the same arguments about underdetermination apply to robustness that apply to proof and refutation? Think of it this way. Suppose I were to say that for any theory which satisfies your rules, there will be indefinitely many incompatible contrary theories which satisfy the robustness requirement equally well.

Pragmatist: You could say that, Quincy, but proving it would be another matter.

Relativist: The proof would be exactly parallel to Quine's proof that for any theory which is empirically adequate there will be indefinitely many rivals to it which are equally adequate empirically.

Pragmatist: Quine proved no such thing. What Quine showed, or came close to showing, was that for any theory with certain empirical consequences, there would be a rival theory with the same consequences.

Relativist: That's what I said.

Pragmatist: I beg to differ. "Having the same empirical consequences" and "being equally adequate empirically" are not necessarily the same thing. And I believe you said that Quine proved that for any theory there were indefinitely many rivals which were *equally adequate empirically*.

Relativist: So I did, but—contrary to what you just alleged—they amount to the same thing. If I have two contrary theories, T_1 and T_2, and they both entail or predict the same phenomena, then they must be equally adequate empirically.

Positivist: I think Quincy is right, you know: same confirming instances, same degree of confirmation.

Pragmatist: I reject your general principle, Rudy, although that is really a matter for another time. What I quarrel with is the suggestion that, in these circumstances, T_1 and T_2 necessarily enjoy the same "confirming instances."

Positivist: But we're supposing in the case in hand that both theories have the *same* empirical consequences, the same positive instances.

Pragmatist: Indeed we are, but why should we suppose that "positive instances" and "confirming instances" boil down to the same thing?

Relativist: And why not?

Pragmatist: I take it that a positive instance of a theory or hypothesis is simply one of its empirical consequences which is true. To call something a *positive* instance of a theory is to remark on the fact that it logically follows from the theory of which it is an instance and that it happens to be true. A *confirming* instance, by contrast, is a special type of positive instance—one which lends empirical support to the theory of which it is an instance. It is my position that *not all positive instances of a theory are confirming instances*. And because that is so, you cannot argue that two theories having the same positive instances are thereby equally well confirmed.

Relativist: You say a confirming instance is a "special type of positive instance," but you haven't told us what type it is. How does one mark the distinction?

Realist: I think that what Percy has in mind goes back to our discussion of testing this morning. You could put it this way: a positive instance of a theory is a confirming instance for that theory only if it results from a *test* of the theory. No test, no confirming instance—although one may have positive instances in the absence of tests. Suppose for instance I propound the hypothesis that drinking coffee cures colds. I then present "evidence" that ten people suffering from colds who drank massive amounts of coffee for twenty-seven days in succession no longer had cold symptoms. Now information such as this, I think we can all agree, is no test whatever of the hypothesis that coffee-drinking cures colds. And for that reason we would rightly hesitate to call that information "confirming evidence," even though the information in question is entailed by the hypothesis in question.

Pragmatist: I actually had a rather different sort of example in mind than Karl's, but the point is substantially the same. Imagine a situation in which, having examined all the cardinals in Hawaii, we propound the hypothesis that "all cardinals have red heads and black bodies," which is in fact a feature of Hawaiian cardinals. Now, if someone challenges our hypothesis and asks us to put it to the test, she will not be much impressed if we rehearse the very information which served as the initial data base for generating the hypothesis to begin with. As we saw this morning, for a test to be a test, it must be drawn from samples different from those used to devise the hypothesis.

Relativist: But that leads to the absurd consequence that one and the same piece of information may be a confirming instance for one theory and no confirmation at all for a second theory, even though both theories imply the evidence in question.

Pragmatist: That appears absurd only because of your knee-jerk disposition to suppose that positive instances and confirming instances must always be the same. Scientists don't think that; statisticians and theorists of experimental design don't think that; ordinary laymen don't think that. Indeed, many of the classical paradoxes of the qualitative theory of confirmation arose precisely from this confused assimilation of positive instances to confirming instances.

Positivist: You're saying that you've got a solution to the paradoxes of confirmation here? I'm afraid that I don't see how that argument would go at all.

Pragmatist: Giving you the full argument would take us too far afield, but I'll be happy to sketch an outline of what the argument might look like.

Positivist: By all means.

Pragmatist: There are, of course, many different paradoxes of confirmation, but perhaps the best known are Hempel's raven paradox and Goodman's grue paradox. They essentially arise in much the same way. I say that because both suppose what has been called the Nicod criterion.

Realist: You mean the principle which says that an observation provides evidence for a hypothesis provided the hypothesis entails a statement of the evidence in question?

Pragmatist: Precisely. The raven paradox is generated by showing that the statement "All ravens are black" is logically equivalent to the claim that "Everything is either black or not a raven." Yet observing a white shoe is a positive instance of the latter statement. And since it seems natural to suppose that anything that confirms a statement confirms anything which is logically equivalent to it, we seem forced to accept that finding white shoes confirms the claim that all ravens are black.

Positivist: Yes, yes—we all know what the paradox is. I was asking how you solved it.

Pragmatist: Well, the foregoing analysis of the raven paradox supposes that anything which is a logical entailment of a hypothesis confirms that hypothesis—that is the Nicod principle. But if we are prepared to reject that principle, by realizing that what a statement entails and what constitutes a test of it are two different things, then the raven paradox goes nowhere.

Relativist: And why not?

Pragmatist: Because we have to ask ourselves whether observing a white shoe seriously tests the claim that all ravens are black. So far as I can see, there is nothing about observing shoes—whatever their color—that would lead us to abandon the claim about the blackness of ravens. Because examining shoes could never in principle refute the raven hypothesis, no observations about shoes constitute a "test" of the raven hypothesis, and—as I have emphasized before—evidence emerges only from tests. As for Goodman's grue paradox. . . .

Realist: Sorry to interrupt, Percy, but I think we are in danger of losing the thread of the argument. Let me see if I can summarize the ground we have recently covered. Quincy was arguing that all theory choices are radically underdetermined. His central argument for that thesis was that there is always the possibility, for any empirically adequate theory, that there are indefinitely many rivals to it which would also be empirically adequate. Percy and I then insisted that *equivalent empirical adequacy* of rival theories is not the same thing as *empirical equivalence* between those theories. We argued that two theories may have the same known empirical consequences yet not be equally well confirmed by those consequences. Quincy challenged the grounds for drawing such a distinction and we unpacked it in terms of the notion of well-testedness. A theory, we insisted, is not necessarily "tested" by all its known positive instances.

Positivist: Let us be blunt about it. Once we realize that 'having the same empirical consequences' and 'being equally well supported by the evidence' are decidedly nonequivalent, then the relativist's alleged proof of the underdetermination of theories by evidence collapses. All the underdeterminationist thesis shows is that numerous theories have the same empirical consequences, and that leaves entirely open the question whether numerous theories will always be equally well supported by the available evidence. It seems to me, then, that the burden falls on Quincy to defend his claim that theory choice is always underdetermined, in view of the distinctions we have been drawing between positive and confirming/supporting instances.

Realist: I think that there is an important argument for realism lurking somewhere here. It has commonly been supposed that the realist's project for trying to find true theories was put in jeopardy by the claim—often associated with Quine—that for any well-tested theory there always will be empirically equivalent alternatives available, at least in principle. This seemed to create a situation in which the realist would be precluded from ever holding that he had good reason to believe that the best-supported theory was true. But what seems clear from this discussion is that there has been a systematic confusion between two very different notions of empirical equivalence.

Positivist: Which are?

Realist: Virtually all the discussions of empirical equivalence in the philosophical literature rest upon the idea that two theories are empirically equivalent just in case they have all the same empirical consequences. Right?

Positivist: Indeed.

Realist: But people have been assuming that if theories are empirically equivalent in that sense—what we might call semantic equivalence—they are necessarily equally well founded or well supported empirically. This latter notion, I hasten to point out, involves *epistemic equivalence*. Now, if we accept Percy's point that positive instances are not necessarily evidencing instances which lend epistemic support to a hypothesis, then we have to say that proofs of the semantic equivalence between two theories—by virtue of showing that they share the same known positive instances—do *not* establish that the two theories are on a par evidentially or epistemically.

Pragmatist: These are intriguing implications that we are tracing out from our distinction between positive instances and confirming instances, but I feel that we are moving the discussion rather far away from its initial focus on relativism.

Perhaps we can ask Quincy if he sees the point of the distinctions we have been trying to draw.

Relativist: I can see the point that there may well be a difference between the positive instances of a theory and its test instances. But even if we focus entirely on the latter, how can you ever be certain that, even if a theory has passed all the allegedly rigorous tests that you subject it to, there will not be indefinitely many contraries of it which will pass an equally demanding battery of tests?

Positivist: Have you doubts that these tests and the rules governing them are rigorous?

Relativist: Of course I do. Indeed, as far as I'm concerned all the rules that Karl and Percy allude to are just conventions for doing science. They have no objective grounding in the facts of the matter and simply serve as convenient instruments for promoting a certain kind of epistemic interest.

Pragmatist: Can we try to deal with one issue at a time? I realize, Quincy, that you relativists are skeptics about there being any nonarbitrary rules for the conduct of rational inquiry and I assure you that we will schedule a future session to allow you to air those doubts. But I think I should remind you that the core thesis that you urged on us today—the thesis of underdetermination—is not a thesis which challenges the objective grounding of our rules of theory selection. Its thrust, rather, is to insist that the rules of scientific method *invariably lead to ambiguous choices*. I wonder if it wouldn't be better to continue playing out that theme before we turn to another?

Relativist: So long as everyone here realizes that I have as many worries about the warrant for your rules as I have about the ambiguity and indeterminateness of your rules, I am happy to accept Percy's suggestion. Let me raise again my question of a moment ago. How, Karl, can you be sure that there will not always be indefinitely many different and contrary theories which will satisfy all the demands your rules of empirical robustness specify?

Realist: I can't be sure, of course. But that is not the relevant question to ask. When I say that the use of the rules of scientific method allows us to pick out a theory which it is reasonable to presume to be true, I am not claiming that the presumption in question is indefeasible. Sometimes our rules pick out theories which subsequently turn out to be false, but that itself is no reason to reject the selection rules. In general, rules of the sort I have described work well at picking out theories which show long-term survival prospects.

Relativist: But doesn't your analysis blithely ignore the fact that important philosophers like Nelson Goodman have shown that even ampliative rules underdetermine theory choice?

Realist: I suppose you're referring to the fact that Goodman once proved that one particular rule of ampliative inference—the so-called straight rule of induction—underdetermined the choice between certain theories.[8]

Relativist: Exactly. He showed that for any hypothesis inductively supported by a certain body of evidence, there were indefinitely many contrary hypotheses which were as well supported by those instances.

Pragmatist: Goodman showed nothing of the sort. What he showed, specifically, was that for any hypothesis, *h,* and a set of its positive instances, *p,* there would be other hypotheses, *h′, h″,* etc., which had the same positive instances.

Relativist: That's what I just said!

Pragmatist: Here we go again. Quincy we've already made the point that positive instances and the confirming instances of a theory or hypothesis are not necessarily the same. Because that's so, one ought not assume that—simply because *h, h′,* and *h″* have the same positive instances—they are thereby equally well supported by the evidence.

Relativist: But Goodman's new paradox of induction showed otherwise.

Realist: No, Quincy. What Goodman showed was that, if one is sloppy enough to suppose that positive instances and confirming instances of a hypothesis are coextensive, then one

8. See Goodman (1955).

ends up in the paradox of assigning equivalent degrees of empirical support to contrary hypotheses.

Pragmatist: I'm not sure we aren't beginning to repeat ourselves.

Positivist: I'm bloody certain we are.

Realist: I agree; may I suggest that we adjourn forthwith, understanding that tomorrow we will tackle the holist issue straight off?

3

Holism

DAY 2, MORNING

Pragmatist: It's clear from yesterday's conversations that the thesis of underdetermination (insofar as it asserts the existence of rival, empirically equivalent theories) may create some embarrassments for the uncompromising scientific realist, but I think that most of us remain unpersuaded, Quincy, that underdetermination does anything to forward your program of establishing that, in principle, one theory is as good as any other. I think the burden falls on you to get things started this morning by giving us whatever reasons you can adduce for that remarkable thesis of cognitive egalitarianism.

Relativist: After reflecting on the matter last evening, I think you are quite right to suggest that the thesis of underdetermination by itself is fairly restricted. In fact, I had been running together in my own mind two related but quite distinct doctrines and using the one label, "underdetermination," to mark them both. In the interests of introducing greater clarity in our efforts at refining relativism, I now propose that we should distinguish sharply between the thesis of underdetermination, which you aptly summed up as the claim that there will always be rival empirically equivalent theories, and the *holist thesis,* which is where we agreed that I should begin this morning.

Realist: So we did.

Relativist: I thought it might provide a useful basis for today's discussion—a kind of agenda setting—if I were to sketch out the central tenets of holism as we relativists see them.

Pragmatist: Fine, but be succinct if you can.

Relativist: Brief it is. Holism implies: 1, when tests are performed, what they test is an entire system of hypotheses rather than a single hypothesis; 2, when a system *passes* a test, we cannot assign a specific level of confidence to the individual component hypotheses of the system; 3, when a system *fails* a test, all we can say for sure is that we made a mistake somewhere—experiment cannot localize the error; 4, given a refuted system—consisting of elements h_1, h_2, . . . , h_n—we cannot determine in advance which elements of the system might be incorporable into a revised system that will be empirically adequate. In principle, any of the elements of the refuted system—save their totality—can be held onto "come what may."

Positivist: That is a large agenda. How shall we begin?

Relativist: The best place to begin is probably with the recognition that holism is a theory of meaning; it teaches that the unit of meaning is not the single term nor even the single statement but rather a whole system of statements, whose terms are interlinked and interrelated in various ways.

Pragmatist: Can you give us an example?

Relativist: If you want a simple example, think of the case of Euclidean geometry. What the term "line" means in that geometry is not fully given by any particular definition within the system. Rather, if you want to know what a Euclidean line is, you have to see how that concept links up with a host of other notions—points, planes—within the corpus of Euclidean ideas.

Positivist: So holism is something like what we used to call the "implicit theory of meaning"?

Relativist: That's surely a part of it, but I want to stress that holism is not simply a theory of meaning. It is also, and perhaps more importantly for our purposes, a theory of testing, i.e., a theory of knowledge. Its epistemic component holds, as I just said, that single hypotheses are never tested in isolation but are always tested as parts of larger complexes or wholes. It thus denies that one can speak, as all of you were glibly doing yesterday, about single hypotheses or theories being "well tested" or "confirmed" or even "corrigibly falsified." Holism insists that it is only rather large systems of hypotheses which are open to empirical scrutiny.

Positivist: Do you mean something like Kuhn's paradigms?

Relativist: A paradigm would be an excellent example of one of these systems. So would a worldview or the doctrines making up a single science. In any event, holism insists that the individual components which make up one of these systems or paradigms can never be directly challenged by experience or observation. At best, all we can hope to discover from experience is that a paradigm or system as a whole has broken down somewhere; error cannot be localized more than that. Indeed, if my fellow relativist Thomas Kuhn is to be believed, we should not even anticipate that we will ever discover that any system as a whole decisively breaks down.[1]

Positivist: Well, which is it to be: can we or can't we discover failure at the systemic level?

Relativist: I think it rather depends on precisely what you mean by "failure." When Quine first formulated this position some thirty years ago, he imagined situations in which an entire system of statements might lead to predictions which were falsified by certain test results. He seemed to regard such outcomes as definite indications of failure somewhere in the system. Kuhn, on the other hand, has quite a different view of anomalies. Like Quine, he readily grants that our "paradigm-induced expectations"—his fancy term for predictions derived from a system—will sometimes be mistaken. But Kuhn argues that such "failures" may easily be attributed to flawed experimental techniques or to a scientist's temporary inability to see how to apply the paradigm in question to the phenomena at hand. In any event, Kuhn holds that such failures, if failures they be, are to be regarded as mere puzzles, challenges to the scientific community to show how apparently refuting instances are actually confirming ones.

Positivist: This is really fuzzy stuff, Quincy. Either a theory or

1. Kuhn writes: "paradigm choice can *never* be unequivocally settled by logic and experiment alone" (1970, 94; my emphasis).

system or paradigm makes a certain prediction or it doesn't. Often, we can ascertain fairly straightforwardly whether the prediction is right or wrong. Surely, when the prediction is badly wrong, that refutes the paradigm or what have you. On these matters, Quine's got it right and Kuhn's just wrong.

Pragmatist: I think that what's needed here is a distinction between paradigms or theories on the one hand and their *versions* on the other. For it doesn't seem to me that Quine and Kuhn are really saying much that is different. I would sketch the distinction as follows: a scientific paradigm consists, for Kuhn, of a set of very general assumptions about what the world is made of and about how the world is to be studied. By themselves, these elements are much too general for the scientist to deduce anything in the way of predictions or explanations. Paradigms have to be fleshed out, filled in with a host of particular assumptions, generated by a process that Kuhn calls the "articulation of the paradigm." Now any particular fleshing out of a paradigm will produce a *version* of it. Over the course of time, one and the same paradigm will thus find exemplification in a variety of successive and even simultaneous versions.

Positivist: I can see where you're going. Your point is that the specific versions of a paradigm are, like Quine's holistic systems, capable as an ensemble of producing predictions which may turn out to falsify them—although not, Quincy will tell us, their specific components. When a version of a paradigm is refuted, scientists know that they must change something in it but the paradigm itself, i.e., the overarching, skeletal assumptions, will be retained. So Quine is saying that paradigm versions or theory versions can be falsified, provided they are understood as broad systems of claims about the world, which is still compatible with Kuhn's claim that paradigms, as opposed to their versions, never directly confront experience.

Relativist: I endorse Percy's and Karl's friendly amendment. Are we agreed then that specific versions of theories and paradigms can be refuted but that the paradigms or theories themselves cannot be?

Positivist: Certainly not. I was helping you formulate holism, not endorsing it. If you want the latter, you'll have to help me see the point by explaining why I should accept the holist analysis.

Relativist: Fair enough; I thought that victory had come too easily! Well, if it's an argument you want, I'm ready to provide it. And I shall draw heavily here on Kuhn's influential analysis of these matters. Basically, scientists approach the world with certain gestalts in their head. These gestalts are interlinked sets of assumptions about how the world is constituted, about the important problems to investigate. . . .

Realist: I thought we all agreed at the first session that the problems a scientist wants to investigate are epistemically irrelevant to theory appraisal.

Relativist: Let me revise that. These gestalts are assumptions about how the world is constituted and assumptions about how the world should be investigated. The former might be called the *ontology* of a paradigm; the latter, its *standards*. Now, before the scientist can do any work with this paradigm—as we just said—he has to flesh it out. This means developing theories about relevant measuring instruments, determining initial and boundary conditions for the design of experiments, and so on. This entire complex is used by the scientist for generating predictions of natural phenomena. When those predictions break down, the scientist will typically hang onto the basic paradigmatic assumptions, i.e., the ontology and standards, and jigger around with the incidentals—what Lakatos once called "the protective belt" of auxiliary assumptions.[2]

Positivist: But what guarantee has the scientist that he can always succeed at preserving his ontology and standards? It is only human nature that he will *attempt* to hang onto those of his assumptions which are the most deeply entrenched in his system of belief. But I take it that you relativists are making a stronger claim than that.

Relativist: Of course we are. We are insisting that there can

2. Lakatos (1978).

never be any evidence or observations which will force the scientist to abandon his standards or his ontology—even if he is behaving in a fully "rational" and "scientific" manner.

Realist: But it is an obvious fact of history that scientists *do* change their ontologies and even their standards.

Relativist: Indeed they do. I don't deny that scientists change their minds from time to time about the most fundamental matters; what I am insisting is that *there is nothing in the evidence which drives that change*. Whether to change or retain his fundamental beliefs is entirely the prerogative of the individual scientist, a fact that David Bloor and Harry Collins have emphasized on numerous occasions. Scientists may give up old paradigms because they are bored with them, or because they no longer have the imagination to see how to bring them into line with experience in heuristically powerful ways. But that reflects nothing about the paradigm itself.

Positivist: You're claiming, then, that evidence didn't force scientists to give up the geocentric paradigm or the Newtonian paradigm or the creationist paradigm?

Relativist: No more than evidence compels us to accept their successors.

Positivist: When you say the evidence neither "compels" us to give up the old paradigm nor to accept the new one, I hope you are not using that term in the sense of logical compulsion. We spent much of yesterday afternoon trying to make it clear that none of us believes that deductive logic alone is sufficient for explaining theory choice in science. When we talk about a choice being "compelled" by the evidence, we mean that the weight of arguments and evidence for a certain choice is ampliatively overwhelming, not that the choice follows with deductive rigor from the evidence.

Relativist: I accept the upshot of yesterday's discussion on that particular topic. When I say that evidence never compels us to abandon the elements that make up the ontology and standards of a paradigm, I mean it in your sense, that there will always be ampliative arguments for holding onto the ingredients of a paradigm as strong as those arguments for giving it up.[3]

Positivist: Thanks for that clarification. I'm glad to see that we're on the same wavelength, at least to that minimal extent.

Realist: What remains true, however, is that Quincy persists in restating his position without offering any arguments in its behalf. I think it's time we asked "Where's the beef?"

Positivist: My sentiments as well. But before we finally try to put Quincy's feet to the fire, I have a brief observation to make. If I understand him rightly, Quincy is about to tell us how the holistic character of theory testing bears out the thesis that the reasons for abandoning a specific assumption can never be better or stronger than the reasons for retaining it. Now what occurred to me was this: if Quincy believes *this* thesis, he must also believe that the reasons for accepting *this thesis* are no stronger than the reasons for rejecting it or for accepting any of its contraries, e.g., the claim that stronger reasons can be given for one paradigm rather than another. Quincy has volunteered to argue in behalf of a thesis which denies that there can ever be even weakly compelling reasons for accepting any belief.

Relativist: And so?

Positivist: Well, I'd have thought the paradox was pretty obvious. If your thesis about the impotence of argument is correct, then it follows that you could not possibly show us by argument that your thesis is preferable to its negation.

Relativist: Why are you positivists so preoccupied with self-referentiality? We relativists can scarcely make an utterance without you trotting out some allegation about its self-referential incoherence.

Positivist: If the shoe fits. . . .

Relativist: I tend to look at matters this way: I believe that I have good reasons for the claim that good reasons can always be given both for and against any paradigm. That thesis, of course, also commits me to holding that there are equally

3. This position is formulated, more or less in these terms, by Doppelt (1978).

good reasons that can be given for the alternative point of view, including yours. . . .

Positivist: May we have that in writing?

Relativist: But you misconstrue my task when you suppose that I am trying to show that the reasons for my position are overwhelmingly weighty. If I will finally be allowed the floor for a minute or two, I will show you that the weight of good reasons never falls on one paradigm to the exclusion of its rivals. That should give pause to all those—including the three of you—who believe that stronger reasons can generally be given for one approach rather than another. It is the *equipotence of reasons* on different sides of rival questions which shows ultimately that reason-giving itself is not, and cannot be, decisive in choices between systems of beliefs.[4] I'm attempting to give you reasons now because that is the game that we four are playing; indeed, I suspect that it's the only game that you three know. But I am doing that only to exhibit in concrete fashion the acute limitations of reasoning and evidence in accounting for theory choice.

Pragmatist: I daresay that none of us finds that very persuasive, Quincy. But rather than pursue this wild hare—which I confess to having introduced into the conversation—let me suggest that you engage in the reason-giving which you have been promising us.

Relativist: Very well. The argument has two parts, one due chiefly to Quine and Duhem, the other to Kuhn, although anticipations of the latter can be found in Quine. The first part of the argument is this: whenever we attempt to bring a particular assumption to bear on nature, in order to test that assumption we must make many other assumptions over and above those ostensibly under test. These assumptions may be about initial and boundary conditions, about the principles underlying our measuring apparatus, and the like. This can

4. Doppelt writes: "the balance of reasons or the demands of scientific rationality [never] unequivocally favors one paradigm (either the old or the new) over its rival" (1978, p. 40).

be represented schematically as an argument having this structure:

$$\{H + A\} \rightarrow O$$
$$\sim O$$[5]

Now obviously from this schema, we can infer that something is wrong in the antecedent which predicted the incorrect result, O; but there is nothing that allows us to infer that the mistake occurred with H rather than A. If we imagine that H is one of our core, paradigmatic assumptions and that A or some part of it (since A will typically be a conjunction of assumptions) is not central to our worldview, then we can always modify A and retain H. That is the force of Quine's claim that any statement we like "may be held true come what may," even in the face of an apparent refutation.

Positivist: You told us that there were *two* parts to the argument. May we have the other before we discuss the first one?

Relativist: Surely. Kuhn has stressed, much more than Quine, that among the elements involved in any paradigm—and in *any* derivation of observational results—are ***methodological standards***. Those standards will involve principles of experimental design, principles for judging theoretical adequacy and the like. Hence he would rewrite the first step in Quine's argument along the following lines: $\{H + S + A\} \rightarrow O$. Kuhn holds that, if the predicted result, O, fails to obtain, then the scientist can always hold onto the components H, which involve some core assumptions about how the world is constituted, and his paradigm-specific standards, S. Any modifications required to bring his picture of the world into line with the observed empirical results, $\sim O$, will be introduced into the auxiliary assumptions, A.

Pragmatist: To put it in a nutshell then, the holist maintains that any substantive assumption about the world can be re-

5. This schema for representing the Duhem-Quine thesis has been canonical since Adolf Grünbaum's early forays against it.

tained in the face of any evidence *and* that any principle about the conduct of inquiry can be similarly retained?

Relativist: That's just about right. I would simply add that the holist also believes it is possible to retain *any particular combination* of H's and S's that one is minded to hang on to. That seemingly minor qualification is in fact rather important since it is linked to Kuhn's idea that the central elements of a paradigm, which will typically include both substantive and methodological claims, can always be retained intact in the face of any conceivable evidence. Well, gentlemen, there you have it: the holist argument for the inability of experience to discredit any particular beliefs we want to hang onto.

Positivist: When you say that "it is possible to retain" any set of assumptions we like in the face of any conceivable evidence that we collect, by making suitable modifications in collateral assumptions, are you claiming that we will always be able to find replacements for A which, when combined with our sacrosanct H and S, will allow us to explain or predict the formerly recalcitrant instances—what I think you called ~O?

Relativist: I'm not sure that a particular scientist will always be able to find a suitable replacement for A—that will depend in part on his ingenuity and his luck. But one will always exist.

Realist: What makes you think so? You're making a very ambitious existence claim; how about giving us a persuasive existence proof?

Relativist: I'll do better than that. I'll show you how, confronted with any apparent anomaly, there will be a modified set of auxiliary assumptions, let's call them A′, which will insulate H and S from refutation. The technique for the construction of A′ is this: certain ingredients of A, either individually or jointly, were responsible for enabling us to forge the inferential link between {H + S}, on the one hand, and the mistaken prediction, 'O', on the other.

Positivist: You mean that if no such elements were needed, then {H + S} alone would have entailed O.

Relativist: Precisely. Suppose that we call that subset of A, needed to establish the inferential link, N. Now, if you want to see a quick and dirty reconciliation of {H + S + A} with ~O, simply drop N out of A. In sum, make our new A′ = {A − N}. With this emendation, {H + S + A′} no longer makes the false prediction and thereby the complex {H + S} has been saved from implication in a refutation.

Positivist: But what a price you have paid, Quincy! Presumably some of the assumptions you dropped out, N, were themselves well-confirmed theories or laws about the phenomena.

Relativist: As usual, Rudy, you fail to see the full implications of the position. Because I am a holist, I deny that any specific theories or laws are well confirmed. The only things that are ever confirmed, or infirmed, by experience are entire webs or clusters of assumptions about the natural world. Because particular claims cannot be tested in isolation, they have no particular degree of confirmation or well-testedness.

Pragmatist: But even if we grant that point for the sake of argument, Rudy is right that you are asking the scientist to pay a very high price when he resorts to your paradigm-saving strategy. Specifically, he has to abandon any claims to be able to explain the sorts of phenomena—to which statements like O and ~O point—for which he formerly expected his paradigm to venture an explanation. Your maneuver would invariably restrict the scope of our paradigms.

Relativist: So it would, but that is simply to say that we sometimes, indeed often, discover that our beliefs about the world were more ambitious and more general than they should have been.

Realist: But the price extracted by Quincy's strategy goes much higher than this. You're not merely saying, Quincy, that observations like ~O are now off-limits to our theoretical inquiries; you're also saying that all those phenomena that we formerly explained by the use of the now jettisoned assumptions, N, are without explanation as well. In short, your strategy requires us to curtail the explanatory scope of our theories. Why would any reasonable scientist favor doing that?

Pragmatist: It's not merely that this gambit restricts the *explanatory scope* of the paradigm—and possibly of paradigms in other fields as well insofar as they depended on some of the assumptions of N to do their work. Quincy's gambit also demands that we accept a paradigm version which is probably going to be ***much less well confirmed*** than its predecessor was.

Relativist: I don't understand that last charge, Percy.

Pragmatist: What I mean is this: it would be most extraordinary, don't you think, if the only use of the assumptions making up the discarded N was to establish a link between {H + S} and the world? Typically the sorts of things that scientists call auxiliary assumptions come into play in a number of applications of a single paradigm and also in applications of other paradigms in other disciplines. For instance, if we're talking about testing an astronomical paradigm, then presumably some of our auxiliary assumptions may involve claims about the behavior of light moving through lenses. Similar assumptions crop up in microscopy, high-energy physics, and a host of other sciences. Now if, in order to save a threatened astronomical paradigm, we decide to jettison without replacement our ideas about how light is transmitted through lenses (and remember, Quincy, that the proposal on the floor is to jettison the auxiliary assumptions without replacement), then we will no longer be able to regard well-tested paradigms in other fields as well tested since the tests which they passed were based on assumptions which you are now requiring us to deny.

Relativist: But the fact remains that a scientist could, if he were sufficiently committed to a particular paradigm, absorb those "costs" and still have a paradigm which was compatible with experience.

Pragmatist: And *there's* the real rub. Scientists are not, and ought not to be, interested in theories which are merely "compatible with experience." If they were, they might as well limit their beliefs to tautologies which are always compatible with any possible experience. Scientists want theories which explain and predict the world, which allow them to manipulate the world in a variety of ways. Theories or paradigms merely compatible with experience are of no use whatsoever. The only strategy that you've described to us for saving a threatened paradigm requires us completely to ignore the fact that scientists seek paradigms with both large empirical scope and with high degrees of empirical support. If the holist position requires us to ignore desiderata as central as those, I for one find holism wholly implausible.

Relativist: You're leaping to conclusions again, Percy. I was describing for you but one strategy for saving a threatened paradigm from refutation. You and Karl have persuaded me that the tactic of rendering a component of a paradigm immune from refutation by abandoning *without replacement* the auxiliaries needed to link it with experience is not particularly desirable. But there are more interesting ways for a scientist to save one of his pet hypotheses.

Positivist: Such as what?

Relativist: Well, what would surely avoid all the problems you have just been raising would be for the scientist who wants to preserve H + S (in the face of {H + S + A} → O and ~O) to develop an alternative A′ which *allows the derivation of the correct result ~O* from the conjunction {H + S + A′}. This technique would involve *no* loss of explanatory scope and *no* loss of supporting evidence, which so worried Rudy and Percy about my initial strategy.

Pragmatist: I have three problems with this suggestion of yours, although I acknowledge that it is a vast improvement on your jettison-without-replacement strategy. The first is not one on which I wish to dwell since we've already aired it in another context. But for the record, we should note that just because your new A′ allows the derivation of the formerly recalcitrant result, it does *not* follow that there is no loss of supporting evidence when we go over to A′. Remember from yesterday that supporting instances and empirical consequences are not the same thing.

Relativist: Duly noted.

Pragmatist: My second problem with your new proposal, Quincy, is this: how do you know that there always will be a suitable A′ which will do the job of establishing an explanatory link between {H + S} and ~O?

Positivist: As I was saying a few minutes ago, we need an existence proof for an existence claim.

Relativist: Surely, if you want a proof of this relatively trivial result, I suspect that we four could patch one together without giving it more than a moment's thought. What you are asking me to show is that there is some auxiliary assumption, A′, such that {H + S + A′} → ~O, even though it's not the case that {H + S} → ~O. . . . I've got it. Simply make A′ = [Q + ({H + S + Q} → ~O)]. Now we have a new auxiliary assumption which plays a central role in establishing an explanatory and predictive link between {H + S} and ~O.

Positivist: This is pretty trivial stuff, Quincy. Your new A′ reeks of the ad hoc. One has only to see it to realize that it was specifically concocted to save an otherwise discredited paradigm version.

Relativist: Of course A′ is ad hoc, but you just asked me for an existence proof and I gave you one. Besides, how could you expect any attempt to save a hypothesis in the face of an apparent anomaly to be anything but "after the fact"? I see nothing pejorative in that. Scientists are all the time trimming their theoretical claims in light of subsequent experience. It's called learning from one's mistakes and you empiricists used to believe in that sort of thing.

Pragmatist: Rudy's point brings me to my third problem, if I may intrude. In accepting your A′ (whether it is this artificial one you have just concocted or some less obviously contrived one), we are also giving up the old auxiliaries, A. Right?

Relativist: Indeed.

Pragmatist: Now, what if those old auxiliaries played an important role in other paradigms or other branches of scientific inquiry? Quite obviously Quincy's new A′ cannot play those roles since its structure is such that *all* it does is to establish a link between H, S, and ~O (via this curious property, Q). Hence, as far as I can see, Quincy's new strategy for saving a threatened paradigm runs the very same risks about loss of explanatory scope and loss of empirical support that we discussed in connection with his earlier jettison-without-replacement model.

Relativist: I cannot give you a proof here and now that there will always be new auxiliaries which will fulfill all the explanatory functions of the older ones and enjoy all the empirical support of the older ones,[6] but all my intuitions about these matters tell me that scientists are generally ingenious and resourceful enough to finds auxiliaries of that sort.[7] Besides, there are plenty of historical examples of scientists who did just that.

Realist: I don't deny that this is sometimes possible, but I totally reject your claim that scientists can always do so. Since you want to talk history, let me ask you: what auxiliaries did the proponents of solid celestial spheres find for explaining the orbits of comets and planets? What auxiliaries has anyone suggested for reviving Aristotle's model of sexual reproduction which will make it the explanatory rival of modern embryology?

Relativist: You're simply confusing the actual with the potential. The fact that no one has propounded an empirically viable version of Aristotle's embryology—and I agree that they have not—is no proof that it couldn't be done.

Positivist: You're trying to tell us that—for all we know—a paradigm of sexual reproduction which holds that offspring derive their traits entirely from the father can, in principle, be reconciled with all the evidence about sperms and eggs, human heredity, and the like?

Relativist: There are no grounds for believing otherwise.

Realist: If we're guilty of confusing the actual and the possible, Quincy, you're muddled over the difference between the *rational* and the *possible*. You claim that because no one has shown that the resurrection of Aristotle's biology is impossible, we should regard "the weight of reasons"—your phrase—as equally balanced between that paradigm and its successors. I think that is crazy. I am prepared to grant that

6. Nor has any other advocate of the Duhem-Quine thesis produced such an argument.

7. Imre Lakatos once put the point this way: "A brilliant school of scholars (backed by a rich society to finance a few well-planned tests) might succeed in pushing any fantastic programme [however 'absurd'] ahead, or, alternatively, if so inclined, in overthrowing any arbitrarily chosen pillar of 'established knowledge'" (in Lakatos and Musgrave, eds., 1970, pp. 187–88).

it's just conceivable that Aristotle's biology might be successfully revived. . . .

Relativist: That's all I need.

Realist: But that is *not* the same thing as saying that, given current evidence, it's as reasonable to believe that human beings reproduce in the Aristotelian fashion as it is to believe that they reproduce in the manner of modern embryology.

Positivist: I think there's some confusion about who carries the burden of proof here. Quincy seems to think it's enough to show that it is in principle possible that any paradigm—no matter how badly it appears to be discredited—could be revived. Karl, by contrast, seems to feel that Quincy must show us that the weight of the current evidence never favors one paradigm over its rival.

Pragmatist: I think that's a sound assessment of the current state of play, Rudy. Relativists are issuing a promissory note on behalf of every paradigm to the effect that, with sufficient effort and ingenuity, it can be transformed into a system which will enjoy all the empirical support of its apparently more successful rivals. The rest of us say that such an outcome seems unlikely. The relativist then says that the burden is on us to show that a discredited paradigm could not possibly be so transformed.

Relativist: That's my challenge to you in a nutshell. And I don't recall that any of you has come even close to producing such a result.

Pragmatist: No, we haven't and there's a good reason for that: it is not clear even in outline form what would be involved in showing that a particular paradigm finally and unequivocally lacks the resources for handling the relevant phenomena, especially if one is allowed to alter the auxiliary assumptions mercilessly.

Relativist: Do I see there a tacit admission that you cannot deliver?

Pragmatist: Absolutely not, for the demand itself is unreasonable. Let's remember how we reached this impasse. You were claiming that the weight of reasons and arguments concerning any two rival paradigms would always be equivalent, i.e., such that neither could claim sufficiently stronger evidence in its favor. Right?

Relativist: Indeed.

Pragmatist: We claimed, by contrast, that certain paradigms were much better supported by the current evidence than their rivals, e.g., modern embryology over Aristotle's embryology. You then conceded that Aristotle's embryology was currently in bad shape empirically but dismissed that as a historical accident. You claimed that, for all we know, a brilliant scientist might come along and revive Aristotle's paradigm to make it equal in empirical support to current ones.[8] Right?

Relativist: Or make it even better. And none of you has shown the impossibility of that.

Pragmatist: Granted, but I now want to convince you that this possibility is a red herring. If the question is—and I take it that this is precisely what the question boils down to—what theory of embryological development should we accept and believe here and now in light of the available evidence, then there is no doubt but that modern embryological theory is to be preferred to Aristotle's. Modern theory has greater explanatory scope, greater precision, and substantially more empirical evidence in its favor. You have conceded all this already, Quincy. The fact is that the current weight of evidence and arguments unequivocally favors modern embryological theory. You have tried to divert our gaze from that conclusion by saying, in effect, "Who knows what Aristotle's biology might be turned into by some scientific genius?"

Positivist: Your point, Percy, if I've got it right, is that no one knows for sure what a revivified Aristotelian embryology might or might not be capable of. But our uncertainty on that score in no way undermines the judgment that, on the existing evidence, Aristotelian embryology has been discredited.

8. Recall Lakatos's remark quoted in note 7 above.

Relativist: But to say that it has been *discredited* is to say that it cannot be revived and you've both granted that you don't know that for sure.

Positivist: I have to say, Quincy, and I make this remark after reflecting on it over the last day and a half, that the problem with you relativists is that you remain infallibilists in a fallibilist age.

Relativist: What nonsense!

Positivist: No, I'm serious. You are telling us that we are entitled to say that a theory has been discredited only if we can prove that it cannot be revived. Like Rorty, you think that the death of foundationalism means the end of epistemology. These are the standards of an infallibilist. I think that the rest of us are prepared to grant that all judgments we make about theories (that this one is well supported, that one not) are defeasible and that they may have to be changed, especially as more evidence comes in. You, however want us to withhold the judgment that one paradigm is really better than another until such time as we can prove that the other is *utterly hopeless*—demonstrably beyond any prospect of redemption. Where we three are willing to let probability be the guide to the scientific life, you're holding us to a much more demanding standard of proof. Since you also think that such proofs cannot be given, you're led to skepticism about all judgments concerning the relative merits of rival theories and paradigms. In a nutshell, I have come to conclude that you relativists are just thwarted foundationalists.

Relativist: That is a richly ironic charge coming from a positivist!

Pragmatist: Can we return to some of the specific arguments about Quincy's holism? It seems to me pretty clear, Quincy, that the holistic thesis that hypotheses can always be insulated from apparent refutation by making changes elsewhere in our system of belief has yet to come to terms with questions of loss of explanatory scope and loss of empirical support; more generally, with questions of rational acceptability. The idea that any statement can be reasonably retained in the face of any evidence is one whose time has come and gone. You relativists have been throwing that slogan around for more than thirty years and you have yet to produce convincing answers to some pretty fundamental questions.

Relativist: That may be so, but the dirty linen can be shared all around, I suspect. If relativists can be charged with having given pretty short shrift to issues of empirical scope and support, it seems to me that the other major theories of scientific knowledge, ably represented by the three of you, have yet to come to terms with the fact that theories and paradigms do not exhibit the patterns of rejection and falsification in which you three firmly believe.

Realist: What do you have in mind?

Relativist: Well, Karl, since you leaped in first, let me begin with you scientific realists and your bizarre claims about refutation. One of your very own, Popper, made the falsifiability of one's beliefs into the touchstone of scientific rationality. Despite having read Duhem, or having claimed to, he pretended through most of his career that falsifications could be decisive.

Realist: That is only a very naive version of Popper's theory of knowledge. As Lakatos among others has shown, there was a much more sophisticated account of scientific inference to be found at the core of Popper's theory.

Relativist: And doesn't this "sophisticated" version of falsificationism tell us that theories are never genuinely falsified? In fact, Lakatos holds that we agree *by convention* to call theories "false"—and he puts scare quotes around the term—after certain sorts of encounters with experience but that we have no genuine basis for believing them to be false.[9]

Positivist: Lakatos says that we should call them "false" when we have no grounds for doing so?

Relativist: Precisely. And if theory falsification just amounts to a (defeasible) social compact to regard certain statements as

9. Lakatos (1978).

discredited and others as not, then I don't see that you realists have any grounds for denying that we can hold onto any statement we like, come what may.

Realist: Popper does say that the decision to regard a theory as falsified is a matter of convention, but it's not an unreasoned convention.

Pragmatist: I must say that the idea of a "reasoned convention" strikes me as thoroughly oxymoronic. If falsifications are conventional, I think it's only words that separate you realists from the relativists. But before you respond to that issue, Karl, I wonder if it might not take us too far afield from our current discussion to deal with the conventional nature of rules of theory assessment?

Positivist: If you're proposing that we take up that issue with care at some future point, I'm agreeable; but if you're suggesting that it's not really worthy of detailed discussion, I would totally disagree. I think that the question of the conventional character of scientific methods may well turn out to be one of the most significant issues before us.

Relativist: I'm with Rudy on this one.

Pragmatist: Right, I've jotted it down for our future agenda.

Relativist: I think that I held the floor when this digression about the conventionalism of rules began. I was suggesting that all three of you have failed to grapple with the problem of falsification and rejection. It's not only Karl and his realist kin who still tend to act as if there are unambiguous falsifications. So do all the positivists and the classic pragmatists: Carnap, Peirce, Schiller, the Vienna Circle—the whole lot.

Positivist: Can you be more specific?

Relativist: You all adhere to a view which says that when a theory or paradigm makes predictions which turn out to be false, then that theory or paradigm should be rejected forthwith.

Pragmatist: And you've yet to show us what's wrong with that position philosophically. . . .

Relativist: Let me rather show you what's wrong with it empirically and historically.

Positivist: I think I get a whiff of another self-referential incoherence about to emerge. Quincy's going to use empirical evidence to refute the hypothesis that hypotheses can be falsified by empirical evidence.

Relativist: That's precisely what I'm going to do, and I've already explained at length why I'm unmoved by your worries about self-referential incoherence. The simple fact of the matter—as Kuhn, Feyerabend, Lakatos and others have shown in repeated cases—is that theories and paradigms always proceed in an ocean of anomalies and apparent refutations. Scientists generally do not, I repeat *do not,* give up their theories when they make wrong predictions. If you continue to ignore that fact, you disclaim any pretense to be talking about the epistemology of *real science*. But there's more to the holist message than this. The holist looks at the history of science and discovers that when theories or paradigms are given up—and of course sometimes they are—they tend to be abandoned in large clumps.[10] Classical mechanics gave way not piecemeal but in its entirety to relativity. Phlogistic chemistry gave way more or less all at once to Lavoisier's new paradigm. And so on.

Pragmatist: Suppose we grant for a moment the historical accuracy of your sweeping claims; what philosophical freight do they carry?

Relativist: Simply this: if you three are right that the component elements of paradigms or theories can be tested individually (and thus accepted or rejected piecemeal), then we should expect paradigms to die a slow death, being slowly whittled away—assumption by assumption—until finally nothing remains. But that's not how it happens. As Kuhn has shown, when a scientific community decides to give up a paradigm, it gives it up in wholesale fashion and entirely; indeed, usually quite abruptly. I claim that to be strong evidence for my idea that the basic unit of acceptance and rejection among scientists is something like the paradigm or

10. Kuhn writes: "the transition between competing paradigms cannot be made a step at a time . . . like the Gestalt-switch, it must occur all at once or not at all" (1970, p. 149).

the broad web of belief, rather than, as you would have it, the single hypothesis or assumption.

Pragmatist: Karl and Rudy may be unmoved by your argument, since neither has ever been much concerned to achieve a close fit between what they say about what scientists should do and what scientists in fact do. But I for one would regard it as quite an important argument for holism if one could show that conceptual change in science has generally been by—how did you put it?—"large clumps."

Relativist: Well, progress at last. I've alluded to only a couple of examples, Percy, but you can look at the writings of Kuhn, Feyerabend, and Lakatos to see plenty of detailed historical examples showing how change of belief in science generally occurs in quite large units. Revolutionaries such as Galileo, Copernicus, Newton, or Darwin introduced wholly new conceptual schemes—more or less full-blown.

Pragmatist: I *have* looked at their work, Quincy, and I'm afraid that I find the historical analysis pretty unimpressive. Indeed, some of the examples you cite (e.g., Copernicus) were scarcely revolutionary at all. More importantly, I've looked at some historical episodes myself and come away from them with the impression that scientific change is much more piecemeal than you or your fellow relativists are generally inclined to allow.[11] We can't begin to settle this issue one way or another here and now, but let me take up one of the examples you cited to indicate why I'm not so impressed as you are by claims about the holistic character of paradigm change. Consider two particularly striking paradigms, the Cartesian and the Newtonian. They were, at the outset, different in all the sorts of ways that you holists get excited about. They involved different notions of space and time (relative v. absolute) and matter (passive v. active). One held that space was full of matter, the other that all the matter in the universe could be compressed "in a nutshell." One allowed bodies to interact only by contact; the other chiefly by forces acting at a distance. . . .

11. See Laudan (1984, chap. 4).

Positivist: We get the point, Percy; the two systems were really very different. Why don't you get to the meat of it and save us the tedium of yet another of your shaggy dog stories.

Pragmatist: The point is this. Despite all those differences in the original formulations of the two paradigms, each was subsequently modified piecemeal in fundamental respects. For example, Huygens—a prominent defender of the Cartesian paradigm—introduced vacuities at the atomic level. Several of the Bernoullis—likewise Cartesians—permitted short-range-distance forces. Among the Newtonians, deep-structure tinkering with the core assumptions of their paradigm was also taking place. Newton himself, for instance, introduced the idea of an all-pervading ether to account for such phenomena as diffraction, cohesion, and capillarity.

Relativist: And so?

Pragmatist: You were earlier claiming, Quincy, that when paradigm change occurs in science, it tends to be all at once and that the holistic character of the change provides support for your view that single hypotheses or theories—especially the fundamental ones—are not assessed individually and rejected or modified in piecemeal fashion. Examples of the sort I have in mind show that all the elements of a paradigm—even the most fundamental and most deeply entrenched—are open to revision, in a one-by-one fashion.

Relativist: You may have shown that some revolutions proceed in a piecemeal fashion, but I don't see that you've shown that all the elements of a paradigm are open to revision.

Pragmatist: I just referred to the case of Huygens introducing vacuities into the Cartesian paradigm.

Relativist: And so?

Pragmatist: Well, I would have thought that if there was ever an apparent case for holding an assumption to be central to a tradition of research, it was the assumption of Cartesian physics that space and matter were coextensive and thus that vacua were impossible. The fact that Huygens and other latter-day Cartesians were willing to modify assumptions as central to their paradigm as that suggests to me that the idea of a totally unassailable hard core associated with every paradigm just doesn't wash. Everything in principle is revisable; but more importantly, that revision often proceeds assumption-by-assumption, rather than by a wholesale and simultaneous modification of the whole package.

Relativist: At best, Percy, you have offered us *one* example of a paradigm that got disassembled one piece at a time. One swallow does not make a summer.

Pragmatist: Of course it doesn't. And I have already managed to bore Rudy to tears. I am content to leave the point this way: there are historical examples of piecemeal paradigm change; there may also be, as you allege, examples of wholesale change (although I seriously doubt it). But I think the historical situation is now sufficiently problematic that you can no longer make the claim you were happy to make a few moments ago to the effect that those of us who rejected epistemological holism were necessarily flying in the face of all the evidence about how science has actually developed.

Positivist: It's not that your stories are boring, Percy; it's rather that I don't see the history of science as doing much philosophical work for us. However, it is salutary if you can show that processes of theory change have often been piecemeal, for that undercuts the Kuhn/Feyerabend thesis to the effect that holistic change is the dominant pattern in the sciences.

Realist: For my part, I urge that we consider temporarily closing the discussion down at this point. A long drink would be more than welcome.

4

Standards of Success

DAY 2, AFTERNOON

Pragmatist: Reflecting on our recent conversations, I sense that we are making some headway with the tasks we have been set. Specifically, it has become clear that the theory-ladenness of observation, the underdetermination of theories by evidence, and the holistic thesis—all key components in contemporary relativist epistemology—do not threaten the claims that, at least in some cases, there are objective rules for selecting between rival theories and that those rules can fully determine theory choices.

Relativist: Lest it look like these sessions are just a relativist roast, it should in fairness be added that it has also become clear that the realist project for finding presumptively true theories about the world is an epistemically quixotic quest, as is the positivist search for theories which can be presumed to be "empirically adequate."

Positivist: Despite your last cheap shot, Quincy, you seem remarkably untroubled by these developments. What's up?

Relativist: The fact is that you believers in scientific rationality have yet to acknowledge the weakest link in the chain of assumptions making up your position. I refer specifically to your conviction that what undergirds the whole scientific enterprise is a set of rules of scientific rationality. As per our agreement at the first session, I have thus far said nothing much about the status of these hallowed rules of yours. I think it's time we turned to that crucial issue. Since you're nominally chairing these sessions, Percy, may I propose that we redirect our attention thusly?

Pragmatist: Karl and Rudy are both nodding their agreement, so I think we can accept your suggestion enthusiastically. If you'd be prepared to kick off the discussion, Quincy, the floor is yours.

Relativist: Gladly. Running through all our previous discussions, as in virtually the whole literature on the epistemology of science, is the presumption that science is, epistemically speaking, a natural kind—clearly demarcated from other forms of belief generation and belief authentication. But what is supposed to mark science off from other doxastic activities? Generally, the answer to that question is: *scientific method*. Scientists purportedly have universal rules for governing their interaction with the world and for shaping their theories about that world; if the rules of scientific method are sound, that would explain what you regard as the empirical success or the robustness of science. To put it briefly, you three believe that science works so well because scientists follow certain objective rules, procedures, or methods for the design of tests and the evaluation of theories.

Realist: Are you about to deny that scientists follow rules or that the rules of science are different from rules used in, say, the pseudo-sciences?

Relativist: I certainly wouldn't deny the first, since one of the most striking features about scientific behavior is how highly regimented it is (e.g., the highly stylized format of the scientific research paper). I probably wouldn't even want to deny the claim that the rules which codify the practices constituting the scientist's "form of life" differ significantly from, say, the rules typically governing religious or artistic experience.

Positivist: So you accept that science is a very different form of belief-generating activity than, say, voodoo?

Relativist: Of course I do. I would deny, however, that science is necessarily a more reliable or more successful way of shaping beliefs than its alternatives;[1] but this line of questioning

1. Feyerabend puts it this way: "An empirical theory (as opposed to a philosophical theory) such as quantum mechanics, or an empirical practice such as modern scientific medicine with its materialistic background, can of course point to numerous positive results, but note that *any view* and *any procedure* that is developed by intelligent human beings has results; the question is whose re-

is not really moving in a constructive direction. Karl and Rudy are acting as if my major concern is with the question of how to *demarcate* science from nonscience. That is *their* preoccupation, especially for positivists like Rudy, but it is *not* mine. For purposes of this discussion, I can happily concede the point that the methods of science may well be very different from methods of justification used in other belief-forming practices. But my central claim will be that those methods are themselves wholly without independent warrant or justification.

Pragmatist: Your worry, I take it, is about the epistemic status of the rules of scientific inquiry. If so, it's a concern that I share.

Relativist: Well, Percy, you're more than welcome to chime in if the going gets sticky. But perhaps I could sketch out my argument a bit before we dissect it.

Pragmatist: By all means.

Relativist: I want to begin with a naive form of the challenge: what is the rationale for the rules of scientific method? and what are those rules? In fact, I should have posed the questions in reverse order, for the answer to the second question underscores the urgency of the first. If I turn to a realist like Popper and ask what the methods of science are, he will give me a certain list; if I turn to positivists like Carnap or van Fraassen or to a pragmatist like Peirce I will get other lists of rules. Now, what's interesting to a participant-observer of the philosophical tribe like myself is that all those lists of rules differ drastically. And all differ from what I might read about "the scientific method" in the opening chapter of an undergraduate physics or biology text.

Positivist: It's true that we disagree about what the rules of scientific inquiry are; but we philosophers differ about all sorts of things. What is your point?

---

sults are better and more important" (1981, pp. 140–41). Elsewhere, he claims specifically that no one has ever shown that science is more successful than other forms of human belief formation: "Nobody has shown that science has results that conform to its own 'wisdom' while other fields have no such results" (1975, p. 32). Peter Winch (1964) has defended a similar view.

Relativist: My point is that there are very different proposals on the table about what the rules of science are and from that it follows that the status of those rules cannot be something like self-evident, analytic truths. If these rules were true or false a priori, we should have been able to settle disputes about them a long time ago. But just as they're not analytic, they equally appear to be nonempirical since none of you go out and do empirical research about what inference patterns scientists actually use, nor would you regard such research as having more than a tangential bearing on your inquiries.

Pragmatist: That's too hasty, Quincy. I for one am convinced that the determination of the appropriate rules for scientific inquiry must be based, in part, on empirical investigation.[2]

Positivist: I'm sorry you said that, Percy, for I had begun to relish the thought that Quincy was the only one of the four of us who routinely propounded self-referential incoherences. But *you* are now suggesting, if I understand it rightly, that we should do *empirical research* to find out what the rules of *empirical research* are? Unless we already know the rules of empirical research, we're hardly in a position to do that; and if we already know those rules, then we don't need to do any research to find out what they are.

Pragmatist: That's exactly the program I'm committed to. And the circularity you allude to is decidedly of the nonvicious variety. Moreover your criticism rests on the mistaken presumption that one can effectively engage in a practice (like doing empirical research) only when the rules codifying that practice have already been made fully explicit. The history of almost any human practice–whether art, science, or music—reveals that an explicit and self-conscious mastery of abstract rules is no precondition for the effective mastery of the practice. If it were, science wouldn't get done at all, since—as I daresay we can all agree—the explicit pronouncements of many scientists, including first-rate ones, about methodological matters are sometimes disconcertingly naive.

2. For a detailed example of such empirical scrutiny of philosophical claims about science, see Donovan et al. (1988).

Relativist: I delight at the prospect of dissension in the ranks, but I wonder if you can postpone your internal squabble long enough for me to set my position out?

Pragmatist: Sorry.

Relativist: Leaving Percy's curious position to one side for the moment, I think I can safely assume that neither Rudy nor Karl believes that the rules of scientific method are either empirical or self-evident. Yes? . . . In that case, what is left? Well, as the realist Popper often puts it, the rules of methodology must be *conventions* which reflect no fact of the matter.[3]

Positivist: Carnap and Reichenbach said much the same thing.[4]

Relativist: To phrase it differently, according to accepted wisdom, methodological rules are neither true nor false; the decision to accept or reject such rules is thus a matter of personal preference or convention.

Realist: Conventional wisdom, as one might say.

Relativist: That pun is too blatant to require comment. My point is that the conventional nature of rules explains, among other things, why scientists and philosophers persist in propounding quite conflicting proposals about the methodological rules and standards that science should obey, and why those debates are never brought to closure.

Pragmatist: But surely not all standards or rules are up for grabs in the way that you suggest. I suppose, for instance, that virtually all scientists—and philosophers of science—would agree about certain standards, e.g., that they want theories that are highly general, testable, fertile in leading to unexpected applications, and generally supported by the available evidence, to mention only the more obvious shared standards.

3. See the opening sections of Popper (1959), where this conventionalist view of scientific methodology is formulated repeatedly.

4. See H. Reichenbach (1938, pp. 10–13); Carnap, for his part, asserted early that norms and values (whether epistemic norms or otherwise) had no "objective validity" since they cannot "be empirically verified or deduced from empirical propositions; indeed, [they] cannot be affirmed at all" (1931, p. 237).

Relativist: There is, I grant you, agreement about the slogans you just rehearsed, and perhaps a few others. But I hope you'll concede in turn that (as Kuhn has shown) the precise interpretation of those demands and the relative weighting of their importance varies from scientist to scientist.[5]

Pragmatist: That's an empirical claim and I don't recall you or Kuhn having produced the relevant evidence to make it convincing.

Relativist: But, Percy, your objection is really beside the point. Suppose it were true, as you claim and I deny, that virtually all scientists and philosophers were agreed about the methods of scientific inquiry. Now, that would be good enough *for me* to say that the rules were locally acceptable, since I think that there is no tribunal higher than the practitioners of a particular practice.[6] But that would cut no ice with the three of you, for you would still want to know whether those methods were correct or appropriate. I alluded to the fact that scientists and philosophers disagree about methodological matters only as a way of directing attention to my central concern, which is that these rules or standards are entirely subjective. Even if *everybody* in the scientific community agreed about them, they would remain subjective since (as Karl, Rudy and I see it) there is no fact of the matter for them to reflect. And how can something be "objective" when there is no fact of the matter for it to reflect?

Positivist: There's another side to the case for the conventionality of rules, a side Quincy has ignored. Methodological rules are imperatives, i.e., "ought" statements. They say things like "You ought to propound testable theories" or "You ought to design experiments in a certain way." We positivists have long held that "ought" expressions cannot be derived from "is" statements. Indeed, instances of such reasoning fall under the naturalistic fallacy, as Percy well knows.

5. Thomas Kuhn develops this view at considerable length in chapter 10 of his *The Essential Tension*.

6. Thomas Kuhn: "As in political revolutions, so in [scientific] paradigm choice—there is no standard higher than the assent of the relevant community" (1970, p. 94).

Pragmatist: I note in passing how strange it is that you positivists happily make common cause with intuitionists like G. E. Moore, who similarly rejects naturalism. But to reply to you directly: I tell my undergraduate logic students about the naturalistic fallacy probably as often as you do, Rudy; but we also both teach them that affirming the consequent is a logical fallacy. That notwithstanding, you are quite willing to advocate the method of hypothetico-deduction as an appropriate form of scientific inference, even though it is a version of affirming the consequent. Why not similarly allow the naturalistic fallacy.?

Positivist: That I teach about hypothetico-deduction, knowing full well that it is a deductive fallacy, is true enough. But the methods that are appropriate for science are not the same as those that are appropriate for philosophy. Scientists routinely make inductive inferences and thus commit deductive fallacies; but I don't expect philosophers to be making inductions and therefore I cannot abide such fallacies within philosophical reasoning.

Pragmatist: Let me make it clear that I'm not endorsing fallacious reasoning in general. But we all know that there are plenty of deductive fallacies which are entirely appropriate forms of ampliative or plausible reasoning. Unlike Rudy, who thinks that the rules of deductive reasoning exhaust the legitimate inferential procedures of the philosopher, I am a naturalist who holds that philosophy can, and should, make use of any of the forms of reasoning appropriate to scientific research.

Realist: I think we are probably getting off the track again. If I remember rightly, Quincy was attempting to argue that the rules of scientific method are subjective conventions which reflect no facts of the matter.

Relativist: Thanks, Karl. To resume: if we once accept that the rules of scientific method reflect no facts of the matter—and I think that both Rudy and Karl agree with me about that—then it quickly becomes clear that science itself is without reliable foundations. After all, these methods—these conventions—are precisely what is supposed to guide the scientist's choices of theory and his interpretations of experiment. If all those choices and interpretations depend on rules which are themselves conventional and which thus enjoy no grounding in the nature of things, then the theory choices based on those rules turn out themselves ultimately to be ungrounded. If rules have no rationale, then choices based on them must be equally irrational.

Realist: That's much further than I'd be willing to go, Quincy. As I see it, we select the methods of science that we do because we believe that those methods will further our epistemic ends. If our belief is correct, then the methods are justified as means to the realization of our ends as inquirers. It is thus the ends or aims of science that determine the appropriate methods of scientific inquiry. Even Popper, whom you accuse of conventionalism about these matters, saw that choices of methodological rules could be justified with reference to the aims of science.

Relativist: But don't you see, Karl, that this maneuver of yours simply generates the problem anew at one remove? For I shall now ask you about the status of "the aims of science" or "the ends of inquiry." There is as much disagreement about those as there is about the methods for getting there. Just among the three of you, I would get wholly different answers to the question: what is the aim of science? That indeed is chiefly what pragmatists, realists and positivists have been debating for the better part of the last century. If our aims are simply subjective (as both realists and positivists generally hold them to be),[7] if our basic goals reflect no facts of the matter, then any methods we propose, and any theories we select or reject because of those methods, will retain a highly arbitrary element.

Positivist: So, is your real argument that the much-heralded "collapse of foundationalism," by removing any a priori

7. Reichenbach calls the choice of the aims of science a "volitional bifurcation," a proposal and a decision rather than a "statement," "and everybody has the right to do what he wants" (1938, p. 11). Popper held similar views about the aims of science.

grounding for basic epistemic values or aims, makes all epistemological and methodological views precarious?

Relativist: That's how one prominent relativist, Rorty,[8] phrases it. My approach tends to be rather more pedestrian than his. I simply ask myself, and you, the following sort of question: if the rules of the game of science (or the aims from which those rules derive) are conventions, then how can you attach any credence to the claim that the theories of modern science have some claim on our doxastic loyalties? Since it is clear that significantly different conventions would produce different candidate theories, a preference for *our* theories, simply because they were picked out by *our* conventions, appears to reveal a remarkable ethnocentrism, i.e., a preference to play the belief-forming game by the rules adopted by our culture.

Realist: Well, as Putnam might say, whose rules would you expect us to play by, if not those of our own culture?

Relativist: I have no objection to your playing the game of science by those rules, *or any others*. What I insist on is the recognition that your rules have no more solid grounding than that they are the rules you have agreed to play by. Other cultures than ours, and many subcultures within our own, have policies quite different from the scientific ones for certifying beliefs.

Pragmatist: You may have taken in Rudy and Karl by your most recent argument, Quincy, but I think that it is rubbish through and through. While they are thinking of some response from their perspectives to your argument, let me tell you how we pragmatists consider these matters.

Relativist: By all means.

Pragmatist: You and I come at these questions from completely different directions. You start from the top down. You say: "Scientists use certain rules and standards for selecting their theories" and you ask "Where do these rules come from?" They're not a priori, you say; they're not empirical, you add. Hence, you conclude, they have no grounding.

8. See especially Part II of Rorty (1979).

And, to underscore your case, you argue that if these rules do have any grounding, then one can raise exactly similar questions about the justification for that. It is the familiar philosophical game of "justificatory ascent" you are playing and I want no part of it. . . .

Relativist: I'm just playing by the usual rules.

Pragmatist: I begin at the other end, looking not at the rules themselves but at the theory choices they have sanctioned. I note the fact that science is a highly effective and successful tool for generating expectations about the natural world. I note, like you, that science also appears to be a rule-governed activity. I say to myself: "Something must be responsible for the striking success of scientific theories; *if* indeed those theories are picked out by certain rules, then there must be something right about the rules in question since a randomly selected set of rules for judging beliefs would not exhibit the striking success shown by the theories of the natural sciences." I'll go further than that. Unless the rules of scientific method reflect something about the "facts of the matter," scientific inquiry would be nothing like as successful as it is.

Positivist: But how can rules, imperatives, ever reflect any fact of the matter?

Pragmatist: I think Karl pointed out a few moments ago that the rules of method are themselves theories or conjectures—conjectures about how objects like us living in a world like this one might sort and choose between the ideas that occur to them. Our theories about inquiry, as embodied in the methodology of science, reflect long efforts by trial and error at figuring out how to produce dependable knowledge.

Positivist: There is a category mistake being made here in Percy's assimilation of rules to ordinary theories or hypotheses. Theories and hypotheses are declarative statements about what there is in the world. They are true or false, and they are *descriptive*. Rules, by contrast, are *normative*, nondeclarative expressions which can be neither true nor false since there is nothing they assert.

Pragmatist: You ask, Rudy, how I can assimilate rules to theories, adding that rules "assert nothing." On the contrary, rules make quite specific assertions to the effect that "doing thus-and-so is likely to produce results of such-and-such a kind." But to show why that is so, I need to set the rules of inquiry within a larger context. Being the creatures we are, we attach a high value to being able to control, predict, and manipulate our environment. There are doubtless plausible evolutionary stories to be told about why we have those values but that issue need not concern us here. Given that we have those values, for whatever reasons, we are looking for ways to fix our ideas, i.e., to find accounts of the world which are generally dependable, readily applicable and capable of anticipating future experience. . . .

Positivist: The problem, of course, is how to identify accounts which will exhibit those virtues; that is the central problem of scientific epistemology.

Pragmatist: After many false starts—some recorded and many presumably lost in prehistory—we have in the last three hundred years been able to produce accounts of the natural world which generally exhibit what we value. Since those accounts have been picked out by certain rules of inquiry, we have good reason to believe that our rules of inquiry are getting better just as (indeed just *because*) the theories they pick out are getting better. You three tell me that the rules of inquiry are just conventions, reflecting no fact of the matter. My response is that there is a fact of the matter which our rules reflect, namely, that inquiry in this particular world works better using these rules than it does using a host of rival rules which humanity has devised for getting at valuable knowledge.

Positivist: You're still missing the central *semantic* difference between rules and theories. Theories are either true or false, even if we don't know which truth value any particular theory has. Agreed?

Pragmatist: Of course.

Positivist: Rules on the other hand have no truth value since they assert no state of affairs which is either true or false.

Pragmatist: I disagree. Methodological rules tell us, in effect, that certain ends can best be achieved by adopting certain means. Any claim about the relation of ends and means involves ipso facto a claim about matters of fact.

Positivist: But it still contains a valuation element, reflected in your term "best." There is a valuational residue here which you are ignoring.

Pragmatist: Let me rephrase it. Methodological rules *presuppose* certain things about the world we live in, e.g., the relative frequency with which the use of certain means leads to certain ends. We can refute wrong methodological rules by showing that the relative frequencies go against them; we can confirm other rules by showing that the relative frequencies support them.

Realist: We're probably all prepared to grant your point, Percy, that empirical information is sometimes germane to the selection of methodological rules. But what Rudy is trying to draw to your attention is that, no matter how extensive your empirical information, there will always remain a nonempirical residue associated with any rule.

Pragmatist: What do you mean precisely by a "nonempirical residue"?

Realist: I mean that, no matter how many facts about the world you knew, you still would not be able to *derive* any rules, not even your conditional ones.

Pragmatist: I grant that point entirely. Rules cannot be *deduced* from empirical evidence, no matter how vast the evidence.

Realist: In that case you've conceded that rules are nonempirical.

Pragmatist: I beg to differ. If the nondeducibility in principle of some belief from the evidence is to count as grounds for regarding the belief as nonempirical, then you'd better start holding that ordinary scientific theories are nonempirical as well.

Positivist: Why? Theories, unlike rules, are declarative statements.

Pragmatist: But you just told us that what made rules nonempirical was their nonderivability from any body of evidence. Theories, too, have that feature—as we all acknowledge. Hume taught us as much. Yet the nondeducibility of theories from evidence does not prompt us to say that theories are *thereby* nonempirical. Why should the nondeducibility of rules from evidence produce agonized handwringing about the status of rules, when a similar nondeducibility of theories produces no such angst?

Positivist: The difference surely is this: if we had a god's-eye perspective and knew all the facts about the universe, then we would be able to deduce the correct theories (indeed, the correct theory would just be the conjunction of all those facts); in the case of rules, however, not even a god's-eye perspective would pick out any rule uniquely. So there is an important difference of principle between rules and theories.

Pragmatist: I'm afraid that I don't go in much for "perfect-being epistemology." We're never going to have all the facts about the world. As a limited creature, with woefully partial access to nature, I propound theories: theories about the world's constitution and theories about how to conduct inquiry. Empirical evidence is relevant to both sorts of theories; we're agreed about that much. Empirical evidence (of the sort we can aspire to) can never do more than provisionally indicate whether we're on the right tract. We agree about that too. Under the circumstances, I don't see that much is gained, and I'm damned sure that much is lost, if we suppose that rules of inquiry—because rules—are just conventions which reflect no facts of the matter.

Relativist: But, even if we lay aside our qualms about whether rules could conceivably be true or false and grant you for the sake of argument that they can be, it's still true that you've given us no philosophical argument as to why we should believe that the rules of scientific methodology are true.

Pragmatist: Since I don't believe that most of our scientific theories are true, strictly speaking, and since I see rules as just another kind of theory, I should hardly be expected to defend the view that philosophy will produce certifiably "true" rules of inquiry. In arguing that we should consider methodological rules and theories on a par, I'm suggesting only that they are on all fours epistemically.

Positivist: But if that doesn't show that rules can be established as true, what's the point?

Pragmatist: I'm claiming that the justification for our theories about the world and for our methods (which, for me, are simply theories about inquiry) are precisely the same: our theories are worthy of acceptance precisely because they work; our rules are worthy of acceptance because they have shown themselves able consistently to pick out theories which work with a pretty high degree of reliability. It's all a matter of finding the right tools for the job. Theories are good tools if they enable us to anticipate and explain natural phenomena; rules are good tools if they lead to the selection of reliable theories. Judgments of truth needn't come into it either at the level of theories or rules.

Relativist: But this idea of what "works" or is "reliable" is itself culturally parochial. Naziism "worked" as a tool for mobilizing German sentiment for the resurgence of German nationalism. Talk to the followers of faith healers and they will tell you that their prayers or charms "work." Doubtless every culture believes that its ideology "works" in some sense or other; otherwise, it would have no rationale for its prevalent ideology. Your suggestion that science (and its associated rules of method) is unique among the systems of belief that human beings have concocted, in that it works, is flatly wrong.

Pragmatist: Quincy, you are confusing two quite distinct issues: whether the holders of a set of beliefs believe that those beliefs are reliable or well-established and whether there are good grounds for holding those beliefs to be reliable. And it's no temporary oversight on your part. The whole thrust of relativist epistemology is to refuse to distinguish between "held beliefs" and "justifiably held beliefs."[9] In your view, there is apparently no form of justification for a belief other than the fact that it is held by a community of believers. By contrast, I hold—and I think I speak here for Rudy and Karl—that the central project for scientific epistemology is to tell us how to distinguish between *justified* and *unjustified*

9. For instance, Bloor and Barnes *define* "knowledge" as "any collectively accepted system of beliefs" (in S. Lukes, ed., 1982, p. 22).

beliefs—regardless of who holds them or whether anyone does. Your position reduces us to saying things like "Jones's beliefs *for him* constitute knowledge; but Smith's beliefs—different from Jones's—constitute knowledge *for him.*" You refuse to acknowledge that terms like "knowledge" carry any evaluational or judgmental element.

Relativist: If I refuse to acknowledge an evaluational element associated with knowledge claims, it is because—in my judgment—epistemology has failed to produce any convincing, non-question-begging way of distinguishing sound beliefs from unsound ones. You traditional epistemologists set yourselves up as judges and arbiters of claims to knowledge when in fact your criteria for making such distinctions are both parochial and arbitrary. *That* is what I meant to underscore when I claimed that the rules of method (which you intend to use to separate "real science" from "pseudo-science") are themselves ungrounded conventions or matters of taste. Those rules make sense to you, and indeed within scientific culture generally, but as soon as you go outside that narrow set of practices and practitioners, you are up against quite different rules for doxastic practice and you have no resources for showing that those rules are any worse or better than your own.

Pragmatist: But, Quincy, I have just been attempting to show you that there *are* ways of justifying the rules of scientific method which transcend the particular practices and interests of the scientific community. I have been claiming that following the methods of science produces theories which confer abilities on us—abilities to control, predict, and manipulate nature—which everyone, whether scientists or not, can see to be in their interests. Your response to my arguments was to say, in effect, that everyone believes his own system of beliefs confers such abilities. My reply, to which you have not responded, is that there is a difference between whether a system of beliefs really confers such abilities and whether such a system is believed by its practitioners to do so.

Realist: This is all a little too abstract for my tastes. I wonder if we can't bring the issue to ground by considering some specific examples?

Pragmatist: If you've a concrete one to propose, I think it would be very welcome.

Realist: Imagine that two groups of people want to build a dam to control flooding along the valley of a nearby river. Suppose that, in order to decide how thick to make the walls of the dam, one group consults the local seer for advice. He reads the appropriate tea leaves and chicken entrails and announces his verdict. The people in his tribe believe in him and thus accept his advice. The second group of people bring in a team of physicists and soil engineers who do their thing, making calculations about water pressure, likelihood of subsidence, and analyze rainfall and water-runoff patterns. This second group takes the advice of their experts and builds a dam to the experts' design, believing it to be well designed. Now, Quincy, I think that Percy's question to you is this: we grant that each group believes the advice it has received. But can't we also say that the one dam is likely to be safer, last longer, and otherwise perform more satisfactorily than the other?

Relativist: Of course you believe that the second dam is better, since you are a member of the relevant tribe in question. So do I, because I also come from a scientific culture. But if you and Percy are asking me, as I think you are, to stand outside my own framework and judge against some neutral standard whether the practices of one culture are better founded than another, I cannot do that and neither can you.

Pragmatist: Whenever we ask you for a comparative judgment, Quincy, you try to turn our question into one which presupposes a foundationalist, Archimedean perspective.

Relativist: In general, you're right about that, although I don't reject all comparative judgments; sometimes we can make comparisons in contexts where the standards required are not in dispute. For instance, deciding whether one thermometer is more accurate than another is a decision taken entirely within a scientific context and there is little room for doubting that the standards for judging goodness in thermometers will be shared by the relevant parties. The cases I object to are those where you are trying to make comparative judgments between the beliefs or practices of communities that do *not* share the same standards. It is in such cases that I deny that there is an objective, non-question-begging procedure for making the choice.

Pragmatist: But the fact of the matter is that people are constantly forced to make just such choices. You seem to suppose that, having grown up in a given culture (or paradigm, to return to one of our earlier locutions), we can never be in a position to decide that rival cultures/paradigms have developed better ways of doing things. And I don't just mean 'better' by the standards of the culture that initially developed the techniques in question. There is virtually no society in the world which takes the attitude toward Western science, technology, and medicine which your relativism would have them take. Large dams these days are built by engineers, period, whether in the first world or the third; science students around the world receive virtually identical training in modern physics and chemistry. No public health official concerned with a tuberculosis epidemic anywhere would completely ignore Western theories of contagion or communicable disease, simply because those theories were not indigenous to her culture.

Relativist: What you're offering is very impressive evidence, not for the objective success of science, but for the long arm of Western cultural imperialism.

Realist: And what you're ignoring is the fact that certain techniques for building bridges, designing airplanes, and treating diseases are demonstrably better than others.

Relativist: But, at the risk of repeating myself, better by *whose* standards, ours or theirs?

Pragmatist: I know that question was intended for Karl, but I'd like to interject my own answer if I may. Indeed, there are three distinct answers I'd be inclined to give to that question, no one of which tells the full story but which collectively may move us in the right direction. The *first* answer is that there are certain human concerns which are universally shared. Health, longevity, access to an adequate food supply, protection from the more extreme ravages of the elements. The universality of such concerns creates a context in which we can plausibly inquire whether certain standards might not be genuinely cross-cultural. If, for instance, a woman wants to ascertain whether she is pregnant (and this is scarcely a concern limited to Western, scientific cultures), she presumably wants an answer which is dependable, i.e., which does not tell her she's pregnant when she isn't and which doesn't tell her she's not pregnant when she is. That standard is surely a perfectly general one. And it is an empirical question whether consulting oracles or running the standard battery of Western pregnancy tests is the more dependable. There is ample evidence that the second is far more reliable than the first.

Relativist: So you and I believe, Percy. But the world is full of millions of people who believe otherwise and who resort to just the practices you were dismissing.

Pragmatist: But believing something does not make it so. There is, I repeat, a fact of the matter here: one of these methods is more dependable than the other and anyone who understands that should act accordingly.

Realist: Remember, Percy, that for Quincy there is no sharp difference between something being so and something being believed to be so. That follows from his refusal to distinguish knowledge (or justified belief) from belief *tout court*. I don't think you're going to make any headway on this particular front; you suggested that you had a couple of other arguments you thought relevant. Maybe you should run those by us.

Pragmatist: Yes, I did. Suppose, just for the sake of argument, that it's true there are no common human standards (like dependability or reliability of predictions); that, in other words, different cultures and different paradigms all rest on different standards for appraising beliefs. Even then, it does not follow—as Quincy seems to think—that practices or ideas can never migrate from culture to culture or paradigm to paradigm sheerly on the strength of their merits.

Relativist: I did not say that ideas never went from one culture to another or from the advocates of one paradigm to another. What I would say was that there could never be evidence or arguments for an idea from one culture so compelling that they could logically force acceptance of that idea in a different culture to which that idea was foreign. And the reason for that is that different cultures, like different paradigms, have fundamentally different standards for the admissibility of ideas.

Pragmatist: I want to concede to you that different cultures may have different standards for, as you put it, the "admissibility of ideas" and yet to show you that, notwithstanding those different standards, ideas from one culture can logically force themselves on other cultures. Let's consider a scientific case first and then try to think up a cultural counterpart. A scientific case might go something like this: suppose that we have two very different groups of scientists, A and B. Group A subscribes to the idea—the standard—that acceptable theories must be mathematically rigorous and generate surprising predictions. Group B, by contrast, holds that acceptable theories must be readily applicable to practical affairs. Would you grant, Quincy, that these are quite different epistemic standards?

Relativist: Yes, sure they are.

Pragmatist: Yet we can surely imagine that some theory may be invented by, say, the scientists in group A which appears to be strong according to *both* sets of standards. In other words, a theory may emerge that is highly quantitative, makes surprising predictions *and* has a host of immediate practical applications. Under those circumstances, we would expect, would we not, that both A scientists and B scientists would accept the theory in question, even though their reasons for doing so would be quite different? Indeed, they would be inconsistent to do otherwise, wouldn't they?

Positivist: I agree with your example, but I don't quite see why it puts the relativist on the spot.

Pragmatist: Central to most forms of contemporary relativism is a commitment to holism. It argues that the standards of a culture (or a paradigm) and the substantive beliefs of that culture (or paradigm) have evolved together and form an inseparable, self-reinforcing package. You can see this bias in Kuhn, Rorty, Wittgenstein, Winch, and a host of others. Now, what the argument I have just sketched out shows is that there is nothing inseparable about standards and substantive beliefs. Accepting the standards of a paradigm or worldview does *not* thereby commit one to holding to a specific set of doctrines about how the world is constituted; on the contrary, one's standards may force one to alter one's substantive beliefs as new evidence comes along.

Realist: Your example shows something else besides, which should give Quincy pause. What it shows is that even if two cultures or paradigms have different evaluative standards, one and the same belief may prove to be dominant in each.

Pragmatist: Precisely. That was the key point of the example.

Relativist: Now I'm thoroughly confused. You were earlier arguing, Percy, that we should conceive of rules or standards as theories about how to conduct inquiry; you're now telling us that having a set of standards does *not* commit one to holding any specific set of doctrines about the world. Which is it to be?

Pragmatist: Your perplexity is entirely my fault and I appreciate your drawing the incoherence to my attention. Let me try to reformulate the point I was just trying to make. It is, of course, true that methodological rules or standards rest on certain claims or theories about how the world is constituted. However, and this is the bottom line, the theories which undergird our methodological doctrines are generally different from the theories which our methodologies provide guidelines for testing. *That* is the sense in which I meant that the methodological components of a paradigm and the substantive part of a paradigm are not inseparably wedded. The rules of inquiry associated with a paradigm will *not* necessarily always give the nod to the substantive theories which have previously been associated with the paradigm.[10] It was

10. This case is argued at length in Laudan (1984).

a failure to recognize this fact that led Kuhn to fall into the error of supposing that paradigms would always be self-reinforcing.[11]

Relativist: That process sounds fine in the abstract, but I note that you've offered no concrete examples.

Pragmatist: If it is examples you want, consider three that come quickly to mind: Cartesian natural philosophers eventually came to accept that Newtonian physics looked better than the physics of Descartes, even by Cartesian standards;[12] Newtonian physicists generally dithered for less than a half-decade after the light-bending observations in concluding that the general theory of relativity was superior to their own; historical research has shown that Planck was led to accept quantum theory not because he subscribed to new standards of empirical appraisal but because he was so scrupulous in applying the traditional standards.[13] In all these cases, the rules associated with a particular paradigm led to the abandonment of the core thesis making up that paradigm. The idea that the standards associated with a tradition or practice always support the core elements of that tradition or practice is simply erroneous.

Realist: And I suppose it was the same supposition that led Winch to argue that substantive disagreements between cultures were incapable of being brought to closure because different cultures had different standards?[14]

Pragmatist: Quite. Moreover, and this is the third challenge I wanted to put to relativism, one of the most serious liabilities of Quincy's approach is that it assumes a rigidity or fixity of standards which is at odds with everything we know about how people (whether individuals or cultures) change their beliefs (and their practices). He supposes that the members of a tribe who are accustomed, say, to taking the seer's advice

11. Kuhn: "each paradigm will be shown to satisfy more or less the criteria that it dictates for itself and to fall short of those dictated by its opponents" (1970, pp. 108–9).

12. See Aiton (1972).

13. See, for instance, Nicholas (1988).

14. For Winch's statement of this argument, see Winch (1964, sec. 2).

about public-works projects have no standard other than, "take the seer's word for it." Thus we discussed earlier a situation in which a Western scientist tries to use the standards of the scientific method to argue that dam building via engineering is more reliable than dam building via entrails-reading. Quincy believes that the members of the tribe will never be *rationally* moved by such arguments since their source of cognitive authority is not the scientific method but the predictions of the seer. Short of the seer himself telling them to use the scientific method, Quincy will doggedly insist that no compelling arguments can be given for their doing so.

Relativist: You saved me the effort of making the point myself.

Pragmatist: But I haven't finished. What is wrong with your point of view is that it ignores the fact that the members of the tribe who want to build the bridge have some overriding concerns. They want a dam which will hold back the water as intended and, presumably, will be no more costly than it has to be. They have hitherto consulted the seer about such matters because, one supposes, they believe that he can give more dependable advice than others. If they can be presented with evidence that the record of seers in designing risky public-works projects is strikingly less impressive than that of modern engineers and scientists, then they will see that it is in their interests to change their standards. It's really only an ends/means argument that is required. If I can show you that you can realize your ends more effectively by using a new means rather than an old and familiar one, you have all the reason you need to change your views about the means.

Positivist: And one might add in passing that it is remarkably patronizing of the relativists to suppose that peoples in other cultures are incapable of, or uninterested in, making calculations about how to realize their ends.

Realist: To speak up on Quincy's behalf briefly, I feel obliged to point out, Percy, that your notion of "more effective" and the corresponding notion held by someone from a quite different culture may well be worlds apart.

Pragmatist: I concede that, but emphasize its irrelevance. I have no doubt that people in different cultures (or in the same culture, come to that) measure social costs and social goods in radically different ways. But however people in the seer's tribe determine such things, we can imagine circumstances in which they may come to the conclusion that certain "alien" practices promote their interests more effectively than long-standing indigenous practices do. And it's not just that they *may* come to such a conclusion; even Quincy grants that much in saying that they may change their minds. Of greater significance is that they will be able to offer compelling reasons for changing their practice—and that process of the reasoned modification of core beliefs, whether of a culture or a scientific paradigm, is something that you and your fellow relativists make no allowances for.

Positivist: I'm not sure what you're driving at by that last remark, Percy.

Pragmatist: What I mean is this: cultural relativists like Wittgenstein or Winch or Rorty and epistemic relativists like Kuhn all suppose that a society (or, in the case of science, a group of normal science practitioners) evolves a set of practices and beliefs which come to be tightly integrated and interdependent. Such a system of beliefs is supposed to be not only tightly interwoven but highly resilient—it is not supposed to be easily or readily displaced. But the relativists tell too clever a story for their own good. For they make the beliefs and practices of a given community so tightly integrated that there could *never* be any compelling grounds for modifying those beliefs. Relativists thus find themselves wholly nonplussed by any major changes in worldview, whether within science or outside it. Having explained to their own satisfaction how well these ingredients hang together, they cannot imagine how anyone could *with good reason* throw out some of the ingredients or introduce new, perhaps radically different, ones into the set. The upshot is that the relativists have no way of explaining how any deep intellectual or cultural change can be the result of reasoning or deliberation.

Relativist: You purport to be describing, among others, the views of Thomas Kuhn. Yet you can scarcely argue that he is caught flat-footed by the phenomenon of broad-scale conceptual change. After all, that is precisely what he wrote *The Structure of Scientific Revolutions* about.

Pragmatist: So he did; and one of the larger mysteries of that book (which is replete with many others) is *why paradigm change should occur at all,* except as a utterly mysterious "conversion experience" or "leap of faith." But it's not just that Kuhn and Wittgenstein have no machinery for explaining large changes like leaping from one paradigm to another. Even more troubling is that they leave no scope for small changes either—at least not where the core components of a culture or paradigm are concerned. Recall that Kuhn is adamant that, once the core ingredients of a paradigm are established, there is to be no tinkering with any of those assumptions; such tinkering would violate what he called the "rules of normal science." For Kuhn, changes—when they occur at all—always occur in wholesale fashion and for inconclusive reasons. As I've already argued before, that's not how conceptual change occurs in science and I suspect that neither is it the way that doxastic shifts in larger cultures take place.

Positivist: At the risk of changing the subject, I think that we have lost sight of one of the key points which Quincy raised earlier in today's discussion. In response to Percy's claim that what justified the methods of science was the fact that they produced successful theories, theories that work well, Quincy claimed that every culture holds that its favorite doctrines "work." Horoscope devotees believe that astrology works; tradition-minded Asians believe that their shamans are able to do things that ordinary folks cannot do. The Azande evidently believe that their poison oracles "work." Within our own culture, many pious Christians believe that petitionary prayer and the laying on of hands "work." If all that is correct, I think that Percy has a problem with his claim that we can sort out the rival standards of diverse cultures, or paradigms, by seeing which ones work and which ones don't. Success, like beauty, may be in the eye of the beholder.

Pragmatist: I think that your point, Rudy, and Quincy's earlier formulation of it, run together some issues which ought to be distinguished. To begin with, I want to make it clear that, although I believe Western science has enjoyed enormous empirical success over the last three centuries, I have no reason to believe that there are not plenty of homegrown practices in other cultures which are also very successful in leading to predictions of, and interventions in, the natural order. Indeed, I should be very surprised if there were not such phenomena in abundance, since my general view of man as inquirer suggests that he is highly motivated to find effective solutions to the technical problems confronting him in the task of getting on with life. So my first point in response to your question is to say that, although portions of Western science and technology are highly successful, I have no grounds to believe them to be unique in that.

Relativist: So you concede that there are ways of producing successful practices without using "science?"

Pragmatist: Of course. To take but two examples, agriculture and mechanical engineering were pretty successfully practiced by a variety of cultures long before there was anything like science around. The issue, for me, however is a *comparative* one: are the methods of science generally more successful at producing what we expect out of a system of knowledge than are the methods of belief-fixation practiced by nonscientific cultures?

Relativist: But, at the risk of sounding like a broken record, we cannot answer that question without reckoning with the fact that different cultures have different standards of what counts as success. Consider Evans-Pritchard's example of Azande beliefs about magic. The Azande believe that, before undertaking an action of any degree of importance, they must consult the poison oracle. To oversimplify, this involves administering a marginally toxic substance to a bird and observing whether the bird survives. Beforehand, it is specified that the death of the bird betokens that one should undertake a particular course of action. Now, here is a case in which *we* would say that the predictions coming from the poison oracle can be no more successful than flipping a coin since, in our view, they stand in no causal relation to the events which will determine whether the contemplated action will go as planned. In sum, *we* would say that the use of the poison oracle is *not* a successful way of deciding what to do. By contrast, an Azande apparently would not consider doing anything important without knowing the prophecy of the poison oracle. He evidently regards it as a quite successful device for determining or influencing the future. My challenge to you, Percy, is to explain how—in a neutral and objective way—you can determine whether our notion of success or the Azande notion should be utilized here?

Pragmatist: The "success" in question depends entirely upon what sorts of events the poison oracle is consulted about and the character of the prophecies it yields. The bulk of the prophecies may concern matters which are not independently ascertainable. Thus one might ask the poison oracle whether a certain member of the tribe is pure of heart or possessed by demons. In such circumstances, one supposes that the oracular pronouncement settles the matter and its success is, as it were, self-authenticating since there may be no independent access to these matters about which the oracle expounds. But if the poison oracle is used to make predictions about matters which can be independently ascertained (e.g., will it rain tomorrow? will there be abundant game at the local watering hole tonight? etc.), then I have to suppose that the Azande are as capable as we are of ascertaining whether the poison oracle has an impressive track record. I daresay that, with respect to events of the latter sort, the oracle will not be strikingly impressive either by our standards or by Azande ones. And if unsuccessful as a predictor of independently ascertainable events, it becomes doubtful as a predictor even of the more hidden states.

Relativist: But you are treating these oracular revelations as if they were straightforward scientific hypotheses to be confirmed or refuted by an examination of their outcomes. That is not how the Azande regard them.[15]

15. This is the position Winch takes on the matter: "Oracular revelations are not treated [by the Azande] as hypotheses and, since their sense derives from the way they are treated in their context, they therefore *are not* hypotheses" (1964; emphasis in original).

Positivist: I think you are being patronizing again, Quincy. If you're right that the Azande do not regard these pronouncements as predictions about whether a certain course of action is risky or dangerous or doomed or whatever, then in what sense can you say that they regard the poison oracle as a *successful* tool for predicting the future? You surely cannot have it both ways. You told us initially that different cultures have different criteria for success and you cited the attitude of the Azande towards their poison oracles as an instance. Percy responded by saying that it would be possible to show the Azande that their oracles were no more successful at making predictions of observable events than flipping a coin was. Your reply was to say that the Azande did not regard the oracular prophecies as anything akin to actual hypotheses about future events and thus they could not be tested in the way Percy proposed. That leaves me wondering just what sort of "success" you, or the Azande, would be willing to attribute to their oracles?

Relativist: To ask the question in that fashion, talking as you do about "success rates" and the like, is to foist a wholly foreign conceptual framework onto the Azande way of thinking about these matters. As Winch has rightly said, "misunderstandings of the sense and purport of institutions like Azande magic arose from insisting on just" such comparisons of Western, scientific notions of rationality with non-Western ones.[16]

Realist: But unless memory fails me, Quincy, it was *you* who first told us that the Azande had a rival notion of empirical success to the one that Percy was proposing. If, as it appears, they consult their poison oracles not because they believe the oracles are successful (either in our sense or theirs, whatever the latter might be), but for some other reason, then your example fails to bear out your own claim that different cultures or paradigms work with fundamentally different views of what constitutes the empirical success of a practice.

Pragmatist: And unless memory fails *me,* we have already gone

16. Winch (1970, p. 251).

past the hour when we reserved a table in the bar. I suggest that we adjourn for now and take up these themes at our next session when, as we have already agreed, the whole issue of incommensurability of rival theories and cultures will form our central topic.

5

Incommensurability

DAY 3, MORNING

Pragmatist: It would be impossible to discuss relativism these days without addressing the vexed question of the "incommensurability" of rival theories, paradigms, and cultures. Indeed, in the writing of such authors as Kuhn, Feyerabend, Rorty, and Quine (the latter of whom discusses the issue under the heading of the "indeterminacy of translation") this has become one of the central slogans of relativism. Yet thus far, Quincy, we've heard almost nothing from you on that score.

Relativist: To set the record straight, I do think that I touched on the question in our opening discussion during the first day. But you're right when you imply that I tend to avoid the term, if only because "incommensurability" has come to have a disconcerting array of quite different senses. Thus, one sense of incommensurability between rival perspectives refers to the fact that the advocates of those perspectives subscribe to *different evaluative standards*. We have already talked through that issue at some length in our discussion. A second, and perhaps more common, sense of the term "incommensurability," raises an issue which we have not really grappled with yet. According to this latter sense (and I propose that this is how we should understand the term in today's discussion) *two bodies of discourse—whether theories, worldviews, paradigms or what have you—are incommensurable if the assertions made in one body of discourse are unintelligible to those utilizing the other.*

Positivist: When you say that the statements in one paradigm are unintelligible to those holding a different paradigm, do you mean that all the statements of the one make no sense in the other or only that some of them fail to have a sense?

Relativist: Relativists have made both claims. Thus, in the early writings of Kuhn and Feyerabend,[1] one finds the claim that no statements within one paradigm make sense in another, what one might call the thesis of global incommensurability. Under pressure from his critics, Kuhn has lately moderated that claim to one of partial incommensurability, meaning that some (but not all) key concepts or statements in one paradigm find no coherent expression in a rival. I myself incline to the view that rival paradigms exhibit *partial incommensurability*.

Positivist: I'm still concerned to get a little clearer about what the thesis of partial incommensurability asserts. Your language thus far has been pretty loose, referring to the failure of certain concepts or statements in one paradigm to make sense to the advocates of rival paradigms. What is this "making sense"?

Realist: You should have no trouble with that one, Rudy, since you positivists have been claiming for decades that many things (e.g., metaphysics) don't "make sense." Moreover, I suspect that Quincy's use of that phrase is probably taken over directly from the positivist lexicon. Quincy supposes, if I've got him right, that rival paradigms or theories have their own associated languages, consisting of terms, and syntactic and semantic rules. When he claims that the advocates of rival paradigms cannot understand one another, he is asserting

1. For instance, Kuhn writes: "the proponents of competing paradigms practice their trade in different worlds. . . . Practicing in different worlds, the two groups of scientists see different things when they look from the same point in the same direction. . . . before they can hope to communicate fully, one group or the other must experience what we have been calling a paradigm shift" (1970, p. 150). Feyerabend: "In short, introducing a new theory involves change of outlook both with respect to the observable and with respect to the unobservable features of the world, *and corresponding changes in the meanings of even the most 'fundamental' terms of the language employed*" (1981, p. 45; emphasis added).

that there are expressions in the language of every theory or paradigm which defy translation (or at least translation without loss) into the languages of its rivals. Correct me if I've got you wrong, Quincy.

Relativist: I'm not sure whether I would prefer to say that the expressions defy translation or that their correct translation can never be ascertained; but for these preliminary purposes the two more or less come to the same thing. So please carry on.

Realist: Now, the relativist is not merely saying that two paradigms *disagree* about certain matters, i.e., that they assign different truth values to certain expressions. . .

Pragmatist: In fact, if I may interrupt, he isn't saying that at all. To the extent that he claims that statements asserted to be true or false in one paradigm find no translation into the language of rival paradigms, he is asserting the impossibility of ascertaining whether rival paradigms agree or disagree about certain matters. Indeed, if they are totally incommensurable, we don't even know whether they are rivals, for rival points of view must be shown to disagree somewhere.

Relativist: What you say, Percy, would be pertinent if I were advocating some form of global or total incommensurability between different paradigms. But as I've already said, I'm not. My view is that some of the statements in any paradigm will be intelligible in alternative paradigms and other expressions will not. We can generally tell whether two paradigms are rivals by focusing on the areas where they are commensurate.

Realist: In any event, I was about to say that the advocate of the incommensurability thesis—forgive me, the partial incommensurability thesis—holds that there are some key expressions of any paradigm which cannot be fully translated into the languages of its rivals.

Relativist: That is precisely the position I hold.

Positivist: Well, we're now reasonably clear about what the thesis asserts. May we have some arguments as to why we should accept this extraordinary point of view?

Relativist: I have been making wagers with myself, Rudy, as to how long we could continue before you would ask such a question. In any event, there are two distinct arguments for the incommensurability thesis, or rather there are arguments for two slightly different versions of the incommensurability thesis. The first comes chiefly from Kuhn and Feyerabend; the second from Quine.

Positivist: Perhaps we can deal with them in turn.

Relativist: Fine. Let me begin with the first. You should find it reasonably congenial, Rudy, since it emerges directly out of the theory of language developed by positivists and logical empiricists. It amounts to the claim that the meaning of a term or concept is given, at least in part, by the network of assumptions with which it is associated. Thus, the meaning of "point" or "line" in Euclidean geometry is different from the meaning of those same terms in Riemannian geometry. The meaning of "space" or "time" similarly varies drastically between their respective Newtonian and relativistic senses.

Realist: This applies alike to the observational and theoretical terms in a theory or only to the more "theoretical" ones?

Relativist: Since we have already agreed that all observations are theory-laden, I reject the presupposition of your question.

Realist: Fair enough; proceed.

Relativist: Well, it's really quite simple from here on. The point is that relativity theory has no terminology for speaking of space and time as they are understood in classical mechanics. We cannot simply take the Newtonian statements about those concepts and translate them directly into relativistic talk of "space" and "time" (as defined in relativity theory); for that would be a transparently inappropriate translation. Nor are there other terms or concepts within relativity theory which will allow us to express the precise sense of classical "space" and "time." Hence, there is partial incommensurability between these two physical theories.

Realist: Your argument about these basic notions seems to be that different conceptual schemes have different languages associated with them and that the nontranslatability of their respective languages follows as a matter of course.

Relativist: That's roughly it.

Realist: But it's surely child's play to work out some translation procedures between any two languages. In the worst case, we can simply *appropriate* the terms and concepts from one conceptual scheme and import them into the other. Thus, we can introduce into the language of relativity theory a new primitive, specifically, the Newtonian conception of space—call it $space_N$. Of course, $space_N$ doesn't mean the same thing as relativistic space, but that should pose no problem. By an extension of this device, we should be able to take any expression in classical mechanics and find its, perhaps coined, relativist counterpart.

Relativist: But what grounds have you for thinking that suitable concepts can be found or conceived, *consistent with the underlying assumptions of the host theory,* into which these terms are being introduced? Suppose, for instance, that the underlying ontology of the new paradigm fails to postulate entities of the sort presumed by its rival?

Realist: Then—as Davidson has argued—we expand its conceptual base, or its vocabulary, until it can accommodate them.

Relativist: But this expansion will presumably require a conceptual scheme to countenance entities or categories which are quite alien to it. At a minimum, it will no longer be the *same* conceptual scheme it was before all these importations. Indeed, I doubt that it will be a coherent conceptual scheme at all.

Realist: Why not? Is English no longer English or no longer coherent because its vocabulary is vastly richer now than it was, say, forty years ago?

Relativist: But English is not a conceptual scheme.

Realist: Yet you earlier invited us to regard conceptual schemes *as* languages. And in any event, if your theory of language is right, it should apply to natural languages like English every bit as much as it does to the language of Newtonian mechanics.

Relativist: We are not merely talking here about the introduction of the odd terms but of fundamentally new and alien concepts.

Realist: I shall stick with my analogy. Is our commonsense conception of the world (which is the framework we Anglophones use English to express) profoundly different because we now have concepts of television and jet engines than it was before?

Relativist: I daresay it is.

Realist: But in that case, Quincy, no two of us have the same conceptual scheme, since concepts and vocabulary differ from one of us to the next. If we were to accept your view on this matter, we should have to say that communication between any two people was wholly impossible. I simply take that to be absurd.

Relativist: I suppose that these things are a matter of degree. What I am struck by is the fact that when rival cultures or paradigms clash, we can always identify a residual incommensurability, a failure of understanding. With your ideas about the ease of translation between rival schemes, we should never encounter cases of noncommunication.

Realist: Think about it this way, Quincy. If I encounter a group of people, scientists or otherwise, who are saying things that persistently resist my efforts to formulate them in my own conceptual scheme, I eventually have to come to the conclusion, not that they have a different conceptual scheme from mine, which defies translation into my own, but that they are not really involved in conversation at all.[2]

Relativist: That dismissive move is altogether parochial, Karl. You seem to suppose in principle that other human beings couldn't have a conceptual scheme genuinely different from yours.

Realist: That's exactly what I'm assuming; or rather, that I could never tell the difference between whether they had a different conceptual scheme or were just making random noises.

Relativist: I don't think that we can possibly sort this one out within the terms of reference that we are using. Rather than attempt to convince you on this issue, I'll propose that we simply agree to disagree. I'd ordinarily be reluctant to drop

2. This is the remarkable position of Donald Davidson (1984).

things so abruptly, but I can see that this particular exchange will get us nowhere and, besides, I have a second argument for incommensurability up my sleeve which I think you will find rather more congenial.

Realist: Let's have it then.

Relativist: Let me ease us into it by asking you to consider an anthropological metaphor that goes something like this: suppose an explorer comes across a hitherto unknown tribe and decides to live among them. No member of the tribe knows his language nor he theirs. No linguist has ever compiled a dictionary of their language. In order to cope, the anthropologist proceeds to learn the natives' language.

Positivist: And how, on your view, does he manage that?

Relativist: Well, he notices what terms they use in various contexts apparently to describe tasks they perform or objects they encounter. By much trial and error, let us suppose, he reaches a point where he can correctly anticipate what sounds they will make in various situations and that he can even carry on conversational exchanges with his hosts. Thus far, this anthropologist has learned a great deal, but chiefly the skill he has acquired is to produce the same noises as his hosts when confronted by the same stimuli. Thus, when he sees an elephant and says "dondu," everyone in the tribe apparently understands him since they nod in assent and say the same thing under similar circumstances. Suppose that such learning experiences are repeated over and again so that a time comes when the natives no longer look askance at the anthropologist's efforts to master their language.

Pragmatist: In sum, he seems to be as fluent at their language as they are.

Relativist: Quite. But, since this is a reflective and thoughtful anthropologist, we may imagine it dawning on him some day that he may have badly misunderstood the natives' language. Thus, when he said "dondu" he meant by that term what "elephant" means in English; but, for all he knows, to the natives "dondu" may mean instead "largest jungle animal descended from the gods." There may well have been nothing in his conversational exchanges with the natives which would determine whether the natives meant the one or the other.

Pragmatist: What you're saying, if I've seen where you're going, Quincy, is that the construction of translations from one language to another is radically *underdetermined* by the experiences of language use. To put it differently, we can never be sure—even if we have attended to native usage very carefully—that *we* have got the real sense of the terms of *their* language.

Relativist: I'd put it more generally than that: there will always in principle be indefinitely many, genuinely different ways of translating any body of discourse from one language to another, translations which preserve the natives' expressions as true, and there will be nothing in the linguistic practices of the speakers of that language that allows us to choose between the translations. But you're right, Percy, to see my argument for incommensurability as a special case of the general argument for the underdetermination of theory choice.

Realist: But I'm still unclear as to what this anthropological parable is supposed to tell us about the relation between rival scientific paradigms.

Relativist: The moral should be obvious, Karl: rival paradigms are different languages for describing and understanding the world. To make what you'd be apt to call a rational choice between paradigms, we need to find wherein they agree and wherein they differ. That, in turn, requires a translation for expressions of one paradigm into the language of the other (or alternatively, translation of both paradigms into some third language—what philosophers like Rudy used to call the "observation language."). But every attempt to translate expressions from the language of one paradigm into the language of another is fraught with precisely the same difficulties that our hypothetical anthropologist ran into when trying to get the "right" version of the natives' language. Specifically, the only behavioral way to find out whether an expression, *x*, in paradigm 1 correctly translates expression, *y*, in paradigm 2 is by seeing whether holders of 1 would say *x* in situations when holders of 2 say *y*. And we have seen that this test of translations is highly inconclusive.

Positivist: You hold it to be inconclusive because. . . .

Relativist: Because speakers of the two languages might have uttered the respective expressions under all the same circumstances even though there may be residual differences in the respective theoretical underpinnings of those expressions.

Positivist: That's what I thought. But then it seems to me that your translation exercise is further compromised by an additional assumption which you haven't yet made fully explicit. I am referring to the fact that your procedure supposes that we must always assume the speakers of a foreign language or a rival paradigm are *speaking truly*. It has us decide on the appropriate translation to assign to native expressions by finding an English equivalent for them which makes them true, if at all possible.

Relativist: Yes, of course. All translation operates with some such principle of charity guiding choices between rival translations.

Positivist: That might be appropriate where translation between natural languages is concerned, but I don't think it will do at all for talking about the comparison of scientific theories.

Relativist: Why ever not?

Positivist: Because such a procedure would do violence to what we already know about the differences between conceptual schemes in science. Suppose, for instance, someone said that in reconstructing Aristotle's physics we should attempt to find translations for his ideas which make as many of them as possible true, where the standard for truth would presumably be what modern-day physicists say. This would produce a terribly anachronistic version of Aristotle's physics and would force us to read his texts in ways that no historian of ideas would find acceptable.

Realist: I think Rudy's right, you know. Take a more germane example: classical mechanics and relativity theory. If one sought to produce a translation of the mechanics of Newton into the language of modern physics on the principle that we should prefer renderings which made Newton's theories as close to modern physics as we could get them—i.e., which made the minimum number of false claims and the maximum number of true ones—then we should find ourselves doing things like translating Newton's claims about "absolute space" into claims about "relativist space," and so on. Surely we don't want to approach the problem of translating rival scientific paradigms by supposing that the best translation is necessarily one which has them both asserting the same things about the world.

Relativist: All that you have said, Rudy and Karl, about potential problems with the principle of charity strikes me as dead right. But I trust you can see that it is further grist for *my* mill. I was earlier trying to argue that, even if we suppose the principle of charity, it is wholly unclear how to translate expressions from one language to another. If we now accept the thrust of your argument, and abandon the principle of charity itself, then translations become even more precarious.

Pragmatist: I wonder if, instead of dealing with this issue in such an abstract way, we shouldn't consider an example or two?

Relativist: Certainly. Suppose that we have a Newtonian and a relativist watching balls colliding on a billiard table. Each will say something like "The red ball changed its velocity on hitting the cushion." Although they utter the same words, the question remains whether they mean the same thing. In fact, of course, the Newtonian—in referring to an altered velocity—will be saying that the ball changed its direction and speed with respect to some absolute frame of reference; the relativity theorist, although using the same expression, will presumably have in mind something like: with respect to some arbitrarily selected frame of reference, the ball changed its speed and direction. They utter the same expression in the same observational circumstances, but that fact obviously fails to establish sameness of meaning for the expressions. Even if we were to find ourselves a situation in which two rival paradigms seemed to give rise to *all* the same expressions about every observable event, we still could not be sure that they are actually asserting the same things about those events. That is the sense in which rival paradigms are incommensurable.

Positivist: If this is the nature of your argument, I cannot for the life of me see why you subscribe to what I think you earlier called a "partial" incommensurability thesis. If your argument here is sound, it would seem to entail a comprehensive and total incommensurability thesis since it holds that we're never in a position to certify that *any* translations between different paradigms are correct.

Relativist: It think I didn't quite follow that. Could we go over it again a bit more slowly?

Positivist: Right. You have told us that in every act of translation between languages, there exists the possibility that the proposed translation fails to capture all that is meant by the statement being translated; in effect, that such tests as we have for authenticating a given translation underdetermine the translation itself. Agreed?

Relativist: Indeed.

Positivist: In that case, the translation of *every* statement from one language to another is suspect, is it not? And, insofar as such suspicions warrant a claim of incommensurability at all, they would seem to justify the claim that different theories are totally incommensurable, not merely partially so.

Pragmatist: It seems to me that there's another twist to Rudy's message. As we earlier noted, one of Quincy's arguments for incommensurability grew out of the implicit theory of definition—the thesis that terms acquire their meaning as a result of the cluster of terms and concepts with which they are interrelated within a given theory. Now, if Quincy's theory of meaning is right, it surely applies to *all* the terms in a theory and not just some of them. And that, in turn, means that either there is total incommensurability between paradigms or no incommensurability at all. Here again, there seems no logical space for some hybrid doctrine of partial incommensurability.

Realist: I'm intrigued by Percy's point, which Quincy quickly accepted, that the thesis of incommensurability is, in effect, a variant of the argument from underdetermination. Most of us agreed, towards the end of our discussion of that topic, that underdetermination poses an epistemological challenge only if one supposes that we require proof that a theory is true before we are entitled to accept it. We said that a fallibilist account of defeasible acceptings and rejectings was untouched by Quincy's argument from underdetermination. It seems to me that an entirely parallel point can be made about translation from paradigm to paradigm. The upshot of Quincy's argument, if I've got it right, is that we can never *prove* that a certain translation of an expression is the correct one, because there remains the ever-present possibility that some different, and quite incompatible, translation would be equally compatible with all the relevant "occasion sentences" uttered by the advocates of a rival paradigm. If we're willing to be fallibilists about theory acceptance and rejection, why not be similarly fallibilist about judging the correctness of translations?

Relativist: I'm quite prepared to be a fallibilist about all matters of belief.

Realist: I don't think you are consistent on this matter, Quincy. For if you're prepared to concede that all we can properly insist on in a proposed translation is the likelihood of it doing justice to the intended sense in the original, then I don't see how you can argue that all proposed translations are on any equal footing. And without that latter claim, the plausibility of incommensurability vanishes.

Relativist: I think I must be missing some crucial step here. Let's rehearse my argument and you tell me where it goes astray.

Realist: Fine.

Relativist: I began by pointing out that the only evidence we have for a proposed translation, *t,* of a phrase or statement, *s,* is that it would be appropriate to utter *t* in all those cases where people have used *s*. Any problem so far?

Realist: No, proceed.

Relativist: Now, it is easy enough for us to imagine that there could be indefinitely many rival translations of *s,* each of them compatible with the occasions on which *s* is uttered.

Positivist: Your imagination is obviously richer than mine!

Relativist: Let me put it differently: it is impossible to disprove that, for any given translation *t* of *s,* there are indefinitely many other translations consistent with all the occasions on which *s* is uttered. That is what establishes the thesis of incommensurability or, as Quine sometimes called it, the indeterminacy of translation. I hope that this is now clear to you, Karl.

Realist: It is indeed; and things are exactly as I had thought. The heart of your argument for incommensurability depends on your claim that none of us has proved that a translation can be unique.

Relativist: And do you deny that you have failed to produce such a proof?

Realist: No, of course I don't. But I remind you, as I started to do a few moments ago, that we are all fallibilists now. Demanding—as you do—that one must be able to disprove a claim before it is reasonable to reject it is just the sort of standard that an infallibilist insists upon. I believe that a proposed translation of a statement can establish its plausibility in the absence of a proof that there is no other translation which captures all the contexts in which the statement is uttered. In the same way, I'm prepared to hold that a theory can establish impressive empirical credentials even in the absence of a proof that no other theory enjoys the same empirical support.

Positivist: Karl's point, Quincy, is that the thesis of incommensurability stands or falls with the thesis of radical underdetermination. Since we've already shown that the underdetermination argument won't hold water, incommensurability suffers a similar fate.

Pragmatist: It seems to me that there is a rather larger context into which this discussion should be fitted. Many contemporary writers, and not the relativists alone, have recently come to the conclusion that the epistemological enterprise is ill conceived. Rorty and others are routinely proclaiming the demise of epistemology.

Relativist: So we are. But what has that to do with the question of incommensurability?

Pragmatist: Well, I may be wrong, but it seems to me that those who want the theorist of knowledge to take down his shingle are suffering from the same preoccupation with infallibility that generates the pseudo-problems of incommensurability and underdetermination.

Relativist: That is rubbish! It was traditional epistemologists, not relativists, who were fixated on infallible grounding and infallible rules of inference. We relativists have simply moved beyond the epistemological enterprise altogether. I deny that there are any indubitable givens, that facts can be directly accessed, that theories or hypotheses can ever be definitively established.

Pragmatist: But in rejecting those doctrines—and you would scarcely expect a pragmatist like me to be among their defenders—you are drawing an inference that I want to resist, namely, that epistemology itself vanishes with the repudiation of incorrigible givens.

Relativist: But what you are seeking, Percy, is still a "first philosophy," albeit one which may be corrigible. We relativists (whether sociologically or psychologically inclined) are *naturalists,* we take science in its own terms and seek no "justification of that knowledge in terms prior to science."[3]

Pragmatist: I find it more than a little ironic that you relativists should describe yourselves as naturalists by way of contrasting yourselves with pragmatists like me. It was, after all, pragmatists like Dewey who were instrumental in this century in articulating what a naturalistic epistemology might look like. But that's probably ad hominem, so let me leave it to one side. The key point to make here is that *traditional* epistemology had three components:

- a search for incorrigible givens from which the rest of knowledge could be derived;
- a commitment to giving advice about how to improve knowledge; and
- the identification of criteria for recognizing when one had a bona fide knowledge claim.

Most of us now agree that the first component must be abandoned, but we continue to stress the importance of the other

3. The phrase is Quine's (1970, p. 2).

two. By contrast, the social epistemologists (e.g., Bloor) and the psychological epistemologists (e.g., Quine, Rorty) seem to think that all three of these traditional concerns must be abandoned with the demise of foundationalism.

Realist: Clearly, only the first of the three is committed to a foundationalist program. I completely agree with you, Percy, that the assumption that epistemic advice-giving presupposes a belief in incorrigible givens is ill conceived. Realists like Popper recognized that long ago when they insisted there was still room for a methodology of science even if there were no indubitable sensory givens. Feeling charitable, I will even concede that several pragmatists have seen the same point. What strikes me as peculiar is that self-avowed pragmatists like Rorty and Quine fail to see it. To say nothing of the sociologists of knowledge who imagine that—since nothing about the natural world is given indubitably—it follows that everything is socially constructed. In the name of "naturalizing" epistemology, the likes of Bloor and Quine would deny epistemology any legitimate normative role.

Relativist: Epistemology once had a normative function because it was thought that it could provide a grounding for science. Now that we know that philosophy is not prior to, nor more sure than, science—and that, above all, is what the naturalistic epistemologist is committed to—we can see that philosophy can provide no firmer grounding for science than science can provide for itself. And that is why normative epistemology has become wholly gratuitous.

Positivist: Tell me about how science "provides a grounding for itself." I should have thought that the whole reason why the four of us do what we do is because we are convinced that science is *not* epistemically self-certifying.

Relativist: The relativist holds that we need to study science empirically in order to understand what Quine has called "the link between observation and science."[4] What he means is that we need to study empirically how we manage, given our sensory irritations—which are our only access to the world—to construct the theories we do, theories which pos-

4. Quine (1969, p. 76).

tulate all sorts of entities which are certainly not "given" in sensory experience. But this task of explaining how we build up our conceptual worlds from sensory ingredients is itself a scientific task; in this case, part of descriptive psychology. We use the methods of science to study the methods of science. And, once we have understood how our knowledge arises from our rudimentary experiences, there is nothing else for the epistemologist to say.[5] The grounding of science is provided when we show how scientists come to believe what they do.

Positivist: But as Rorty points out at great length, this is to confuse explaining a belief with justifying a belief.

Pragmatist: And what about giving advice on how best to construct theories? What about telling us when it is reasonable to regard a theory as well supported by the data? What about telling us when it is legitimate to prefer one theory to a rival? These remain important epistemic questions, even in these postfoundationalist times.

Relativist: I'm sorry to differ with you, Percy, but those questions—insofar as they are legitimate at all—are simply scientific questions. Scientists have to answer them all the time since they are routinely involved in processes of theory assessment and theory construction. They do not need the help of philosophers in that regard.

Realist: Whatever makes you think that? Are you supposing that scientists never make mistakes in judging the merits of rival theories or in interpreting data or designing experiments?

Relativist: The only category of "mistake" that I recognize has to do with failures of consistency or internal coherence. I understand what it means to charge a scientist with making a mistake in designing an experiment, *if* by that one means simply that the experiment was conducted in a fashion that violates customary practices in science. I similarly under-

5. Quine: "Epistemology, for me, or what comes closest to it, is the study of how we animals can have contrived . . . science, given just that sketchy neural input" (Quine, 1981, p. 21).

stand what it means to claim that a scientist made an erroneous theory choice *if* that means a choice out of line with the principles of theory choice then prevailing in the scientific community. But I don't think that this is what either Percy or Karl has in mind when *they* talk about the normative functions of epistemology. I think that they want an *independent tribunal* for assessing the codes and norms implicit in scientific practice. But that sort of independence could be obtained only if there were a source of knowledge which did not itself presuppose science. I deny that any such "first philosophy" exists. So do the rest of you, come to that. The difference between us is that I am honest enough to admit that science is all we've got *and* therefore that there is no higher tribunal. The rest of you fundamentally agree that science is the only ticket to knowledge in our culture but still hanker after some way of showing that science is justified with respect to some elusive higher standard. You haven't realized that those two views simply don't fit together.

Pragmatist: I do want some way to appraise the norms of science, for otherwise their invocation is arbitrary; one might as well consult a seer or read tea leaves to decide what to believe. But you have me wrong when you suppose that I expect to justify these norms by going to some higher discipline or first philosophy. In my view, we appraise the norms of science by asking whether they work, i.e., whether they lead to effective anticipations of, and interventions in, the natural world.

Relativist: But how do you decide whether they "work"? That is surely a question of whether those norms are empirically well supported. But how do you decide what empirical support is? By invoking the norms of science, which are constitutive of what we mean by empirical support. That is question-begging if anything is.

Pragmatist: Your analysis is too hasty by half. Take a concrete example. Suppose I want to appraise the rule "Always prefer theories which postulate fewer entities to those which postulate many." How might I evaluate this rule? One relevant thing I might try to find out is whether Ockham-like theories have stood up to subsequent testing better than theories which were ontologically bloated. Let us suppose, as I strongly suspect might be the case, that theories postulating multiple entities have generally fared better than theories postulating a bare minimum. Under such circumstances, I would be entitled to say that Ockham's razor gives bad advice and should be excised from scientific epistemology. Where, Quincy, is the circularity here? I haven't used Ockham's razor (or any of its contraries) to evaluate this rule; yet I have scrutinized it empirically. This is the sense in which, as far as I'm concerned, epistemology can have a normative role.

Relativist: What you are doing in this case is using certain rules of science (e.g., the rule that we want theories to stand up to future testing) to assess another rule—the principle of economy. What your hypothetical example would show is that those rules are not internally coherent. But that provides no license for saying that Ockham's rule is wrong. We could equally well hang onto Ockham's razor and throw out the rule which says that theories must stand up to further testing.

Pragmatist: You can't be serious, Quincy. The principle that our theories should stand up to further tests is not merely, as you call it, a "rule of scientific method." It is one of the central aims of all inquiry. To abandon that principle would be to say that we are quite indifferent to whether our theories work or don't work; and that would be to misunderstand why we develop theories to begin with.

Positivist: This lengthy digression about the credentials of epistemology is interesting but it has, I believe, taken us well away from this morning's central topic. I wonder if there is anything more to say about incommensurability before we break for lunch?

Pragmatist: I'm happy to take up the cudgels here because, all through our earlier discussion of the topic, I got the impression that you three were regarding incommensurability as much more of a threat than it seems to me to be.

Positivist: But Karl and I were attempting to show that the threat from incommensurability can be defused.

Pragmatist: I understand that, and I agree with the upshot of your challenge. But it seems to me that, even if rival theories were partially incommensurate (and I think that Quincy has yet to establish that result), it is not clear to me why that would threaten scientific rationality. Specifically, it seems to me that the explosive potential of incommensurability can be quickly defused by pointing out that the *wholesale translation* of the claims of one paradigm into the language of its rivals *is not required to make rational choice between those rivals*. Insisting on "translations between paradigms" and "sameness of meanings" for their respective expressions—demands which give rise to the problem of incommensurability to begin with—has never struck me as essential to the process of choosing between rival theories. I wonder if we might not, solely for the sake of the argument, grant Quincy his point that interparadigmatic translations are frequently underdetermined and yet still maintain that this provides no ammunition for the relativist thesis that interparadigmatic choice is thereby rendered nonrational?

Positivist: I think that you'll give the game away if you proceed in that fashion. For one thing, unless we can establish translational linkages between rival paradigms, then we don't even know whether they make conflicting predictions; and if we can't ascertain that, it is hopeless to attempt to design any tests which would enable us to choose between them.

Pragmatist: Just hear me out. Let us suppose, as Quincy invited us to suppose, that two paradigms, A and B, are *partially* incommensurable, i.e., that there are some claims made by each which are not translatable into the language of the other. However, since A and B are only partially incommensurable, they are also partially commensurable, i.e., there are some claims made by B which are formulable within A and vice versa. Hence by focusing on the discrepant but commensurable claims made by A and B, we should find plenty of tests to use for choosing between them.

Positivist: But I've just shown that Quincy's analysis commits him to total incommensurability between rivals.

Pragmatist: So you have, and I agree; but in the interests of clarity, it still seems worthwhile to ask whether some partial form of incommensurability between rival perspectives—should it exist—would rule out the possibility of rational choice.

Realist: It seems to me, Percy, that your argument of a minute ago presupposes that the area of commensurability between the two paradigms has to do with their observable implications. Couldn't there be situations in which two paradigms were partially commensurable but their areas of commensurability dealt with matters about which we could not perform experiments?

Pragmatist: In a word, no. The advocates of partial incommensurability such as Kuhn and Feyerabend invariably base their claim for that thesis on supposed failures of translation at the highly theoretical level. Moreover, if two paradigms were fully incommensurable at the observational level, then—as we have already noted—it would be impossible to regard them as rivals between which choice was necessary. So, to repeat, if we have two genuinely different, but partially commensurable, paradigms, there must be some common empirical phenomena which they both address. It is there that we focus in designing tests to choose between them.

Positivist: Such as what?

Pragmatist: For one thing, we could examine some of the predictions made by each. Suppose we discover that some of the predictions made by one of the two turn out to be false and that we have discovered no refutations of the other. Under those circumstances, we would be warranted in rejecting the one and holding onto the unfalsified one.

Relativist: Why must I keep reminding you that there are no falsifying experiments in science?

Realist: You seem to forget, Quincy, that we have already covered that ground pretty thoroughly and the three of us are convinced—I believe we even won your concurrence—that there are circumstances in which we are entitled to suppose that a certain theory or paradigm is false. What interests me about Percy's point is that I wonder whether there are any *positive* tests that could be performed to choose between partially incommensurable paradigms. It is one thing to learn that a paradigm, or paradigm version, is false; but I'm interested in exploring the question whether there are circumstances in which we might be able to reach a positive verdict. Percy glibly said that if one paradigm is falsified, then we could hold onto its rival; but that is a pretty pathetic rationale for acceptance.

Pragmatist: Your point is fair, and I think that there is an answer to it. Suppose that one of our two paradigms makes some surprising predictions successfully and its rival utterly fails to predict anything very surprising, or makes surprising predictions that are false. Wouldn't that then be positive grounds for preferring the former?

Realist: Of course.

Relativist: Hang on a minute. How do you decide whether the predictions made by a paradigm are *surprising*? Surely, the paradigm that makes the predictions won't label its own predictions "surprising"; on the contrary, within the frame of reference of that paradigm, anything it predicts is wholly *un*surprising. If the notion of surprising predictions is to be coherent at all, it must refer to the fact that a paradigm (or rather a specific version of a paradigm) makes predictions which are, according to rival paradigms, quite unexpected. Right?

Pragmatist: I suppose that's right.

Relativist: But if the paradigms in question are incommensurable—and you've supposedly been granting me that point at least for this discussion—then you can't tell whether the predictions made by one paradigm are surprising vis-à-vis its rivals because you can't do the relevant translations to ascertain its surprise factor.

Pragmatist: That's a good point, but not necessarily a decisive one. What I have been granting you, and all you claimed at the outset of today's conversations, was that rival paradigms are *partially* incommensurable, i.e., that there will be some statements in any paradigm which cannot be rendered coherently in the languages of its rivals. Since the incommensurability is only partial, that means some statements within each paradigm *can* be rendered into its rivals. Hence, whether we can determine that any specific prediction made by a paradigm is surprising depends upon whether that prediction can be translated into the language of a rival paradigm and, once translated, whether the rival paradigm attaches a high likelihood to the predicted outcome.

Relativist: Fair enough, Percy, but you've no guarantee that any particular prediction will be translatable—precisely because the two paradigms are partially incommensurable.

Realist: Big deal. That is simply to say that we're in no position to decide about some of the predictions made by a paradigm, whether they're surprising or not. But that leaves completely open the possibility that other predictions will fall squarely in the category of surprising predictions and, if confirmed, will constitute strong evidence for the paradigm version that makes them.

Positivist: Having lent Karl and Percy a helping hand on this issue, I do have a problem with the great stock they set by the ability of a theory or paradigm to make successful, surprising predictions. Specifically, it arises out of the way in which you two allowed Quincy to define how we determine whether a prediction is surprising. He proposed, and you agreed, that the prediction made by a paradigm is surprising just in case it is a prediction assigned a low probability by a rival paradigm. Right?

Realist: Sure; what's the problem?

Positivist: Well, that seems to make the empirical support for a theory or paradigm radically contingent on the historical accident of the rivals with which it is being compared. There presumably are conceivable rivals which would make a certain phenomenon, *p*, predicted by a particular paradigm, "surprising" and other rivals which would make *p* quite unexceptional. How can we claim to be making "objective" appraisals of a theory when those appraisals depend on the accident of which theory or theories immediately preceded it?

Realist: I think it all goes back to the notion of a "test" which we were kicking around the other day. The reason why we prefer theories which have successfully made surprising predictions has nothing to do with the surprise factor per se. As you have all rightly pointed out, what we mean when we call a prediction surprising is that it's a prediction which we would be disinclined to make if we were using one of the rivals to the theory under consideration. By seeing whether the surprising prediction is correct, what we are actually doing is designing a crucial experiment between two theories or paradigms, one of which makes a certain prediction (what we call the surprising prediction) and one of which does not. This is a good test because it provides us specific grounds for choosing between rival approaches.

Relativist: But Rudy's question remains unanswered. I repeat: doesn't this make our assessment of a theory or paradigm hostage to the theories or paradigms with which it is in active competition? And isn't that largely a matter of historical accident?

Pragmatist: Karl and I seem to be of one mind on this issue so I'd like to intervene on Karl's behalf. It does, of course, make our assessment hostage to the competitive state of play in science at any given time. But that's just to emphasize what should have been clear all along—that *assessment in science is invariably comparative*. When we say that a theory has been well tested, we don't mean—or at least we shouldn't mean—that it has been tested and picked out as superior to all possible contenders. What we mean rather is that the theory in question has passed tests more impressively than any of its known rivals, and that therefore it should be preferred to its known rivals.

Relativist: But if all we are learning from an experiment is that one theory is to be preferred to its known rivals, that puts us in no position—as Karl would have it—to conclude that the preferred theory is true or likely to be true.

Pragmatist: I think that's quite right. When I said that I agreed with the realist on these questions, I was not referring to his optimistic assumption that we can reasonably conclude that a theory is true by virtue of its passing certain tests; I was rather indicating my agreement with his thesis that the "surprising" feature of a prediction is of epistemic rather than merely psychological relevance. Incidentally, precisely the same difficulty facing Karl confronts Rudy's notion of empirical adequacy.

Positivist: How so?

Pragmatist: It is surely clear that the ability of a theory to make correct surprising predictions—i.e., to make predictions to which its known rivals assign a low probability—is no telling indicator of the theory's long-term empirical adequacy.

Positivist: And why not?

Pragmatist: For one thing, when we say of a certain theory that it is empirically adequate, we are making a very ambitious claim on its behalf; we are saying that there will never be observations that refute it. Since our track record at finding theories which will stand up well indefinitely in the face of future tests is not very impressive, one should be very wary about supposing that any theory—whatever empirical successes it can lay claim to—is likely to stand up to extensions of it to new domains.

Positivist: But claiming that a theory is empirically adequate is much less risky than claiming that it is true.

Pragmatist: Yes, but that's rather like the blackjack addict saying that playing blackjack is less risky than feeding slot machines. Both are very likely to lose you money. In the same fashion, both truth judgments and empirical adequacy judgments are likely to lead you astray.

Realist: But what's your pathetic alternative, Percy? All you would have us infer from some series of tests is that one theory passed them and that certain other theories failed them. That tells us nothing important epistemologically about how the theories stand.

Pragmatist: That is my alternative, Karl. But I submit that it tells us everything we really need to know. It tells us that, among the theories we have been able to conceive, one theory has passed more demanding tests than its rivals and is thus likely to stand up better to future scrutiny than those rivals will.

Positivist: Do I detect the notion of empirical adequacy creeping in here?

Pragmatist: Not at all. In judging that one theory is more likely to stand up to future testing and applications than another, I am making no assumptions about whether the preferred theory will stand up successfully to *all* future tests. In fact, I would be very dubious of any such inference. And that is what empirical adequacy requires. Theory choice, as I see it, is always choice among extant rivals. Learning which among a set of known rivals has stood up best in the face of demanding tests is the single most important thing for a theoretical scientist to learn.

Positivist: I don't know about the rest of you, but I'm beginning to think that lunch should be preferred to all its known rivals, including the prolongation of this particular conversation. May I suggest that we adjourn?

6

Interests and the Social Determinants of Belief

DAY 3, AFTERNOON

Pragmatist: We're coming to the end of our time together and I'm concerned that so far we have been focusing chiefly on the negative side of relativism, i.e., its denial that evidence is decisive in shaping scientists' loyalties, its insistence that the weight of reasons never comes down squarely on one paradigm rather than another, its belief that science never makes any form of nonlocal progress. I don't think that Karl, Rudy, or I have been persuaded that the case for relativism is particularly impressive on those fronts. At a minimum, the jury is still out where those theses are concerned. But I wonder if it mightn't help us to get a more rounded sense of the relativist position if we were to explore the positive tenets of relativism.

Realist: It's not clear to me that relativism has *any* positive tenets. It seems to be chiefly a form of skepticism, and the skeptic notoriously has no positive claims to make about the knowledge-getting enterprise; or rather, the skeptic makes positive claims on pain of self-indictment.

Relativist: I have said repeatedly that relativism is *not* a form of skepticism. If the rest of you are so inclined, I'm more than willing to sketch out what we relativists think about scientific belief.

Positivist: I'm certainly game.

Pragmatist: Let me suggest a context which might be a useful one from which to begin our discussions, and Quincy can then take it from there. For all the differences among the four of us, there are several central points about which we are agreed.

Positivist: There are? This I've got to hear.

Pragmatist: We're all agreed that theories play an important role in the scientific enterprise, that scientists from time to time change their views about what the central theories of their discipline should be and that observation and deductive inference by themselves are insufficient tools to permit the scientist to decide which theories he should accept. We've also agreed that modern science confers an impressive degree of predictive and manipulative control over nature (although Quincy has expressed doubts that there is anything unique to science about that). My basic question for Quincy really boils down to this: since you don't believe that evidence and argument are the engines which drive science, since you don't believe that scientists ever have compelling reasons for changing their theories, but since you also believe that scientists *do* change their theoretical loyalties, I wonder what alternative story you have to tell about those doxastic shifts?

Relativist: I welcome that context in which to elaborate my position. What I take seriously, and you three ignore, is the elementary fact that science is a *social* and a *human* activity. Science doesn't take place in some disinterested Platonic world of mental relations; it is produced by scientists who have all the same interests and self-concerns that ordinary folks do. You three invariably fall into thinking that science is the realm of disembodied ideas.

Positivist: If that is a risk we run, you are sailing perilously close to treating science as a set of lobotomized social institutions, where everything matters except ideas!

Pragmatist: Gentlemen, trading epithets will get us nowhere. We agreed, Rudy, to give Quincy a chance to set before us what positive theory of scientific belief he has. I think we should allow him to proceed.

Relativist: As I was trying to say, scientists are people with careers to build, reputations to preserve or enhance, influence to acquire or extend, egos to protect, and skeletons to hide.

Apart from their own personal interests, they also have larger loyalties to their society, their class, their religion, their race, and their gender—most of which act at a less explicit level. If you really want to know the kind of story that I'm inclined to tell about why scientists adopt the theories that they do, the stories will be compounded of ingredients such as the ones I have just mentioned.

Positivist: This is all still very . . . .

Relativist: Very vague; yes, Percy, I know it is but give me a chance to fill it in a little bit. Imagine the case of a senior scientist whose whole career has been tied up with the fortunes of a particular theory, a theory he may even have been responsible for formulating in the first place. That theory, let us suppose, is what he is best known for; it has brought him an impressive string of research grants, postdoctoral students, and research awards. Suppose eventually that some younger scientist somewhere produces evidence that he construes as refuting the theory of the senior scientist. Does anyone of us really expect that the senior scientist will be able to give a dispassionate and disinterested assessment of his theory under those circumstances? On the contrary, we find it entirely predictable that he will squirm and maneuver, desperately seeking some way to preserve his theory in the face of this threat to his cognitive authority within the scientific community.

Realist: You're claiming, if I may generalize, that scientists have what one might call their "professional interests" which are often at odds with the disinterested pursuit of knowledge for its own sake.

Relativist: That's precisely what I'd say, although I'd probably tend to make it even more general than your formulation suggests. In my view, scientists have interests not only in their professional standing as scientists but also in terms of their other roles and concerns, ranging from the economic to the political and religious.[1] Thus, a deeply Christian scientist

1. Compare David Bloor: "there is much evidence that features of culture which usually count as nonscientific greatly influence both the creation and the evaluation of scientific theories and findings" (1976, p. 3).

like Newton is going to have no truck with a physical theory which he perceives to be undermining the Christian message. A contemporary biologist with Marxist political orientations is not likely to be sympathetic to a theory like sociobiology which holds that most human social traits are acquired genetically. A scientist like Einstein who has a priori metaphysical convictions about the deterministic and causal structure of experience is never going to accept quantum theory as the last word.[2]

Positivist: Judging by the tone in your voice, you expect all this to come as a revelation to the rest of us, as if we were totally unaware of the influence of such factors on scientists.

Relativist: I wouldn't exactly accuse you of being unaware of these factors; but your way of acknowledging their existence is to regard them as aberrations, occasional and unfortunate departures from the scientific ideal. I, on the other hand, see them as the scientific norm; none of you has begun to come to terms epistemologically with their omnipresence. You still see the scientist qua scientist as a one-dimensional figure, disinterestedly and exclusively committed to having experience guide his beliefs, and to letting the theoretical chips fall where they may. That picture, if I may say, is pure mythology, designed to lend a semblance of objectivity to an activity which is largely driven by noncognitive concerns of power, prestige, and influence.

Realist: You won't have to be told that those are fighting words, of course. Before the battle itself begins, let me do some skirmishing around the edges. You began by giving us some examples of scientists—Einstein and Newton—who had strong ideological axes of one sort or another to grind. I have two questions to ask about that. The first one is simply this: how far do you think it will take you in explaining a scientist's beliefs to invoke her theological and metaphysical views?

Relativist: I'm not sure what you're asking.

Realist: What I mean is this. Suppose I grant you that Einstein's opposition to quantum mechanics was based chiefly

2. See Fine (1987).

on some of his a priori convictions about the nature of causation.[3] Still, there is much important science he did—from the development of the special theory of relativity to studies on the photoelectric effect—which appears to have been completely uninfluenced by his metaphysics of causation. Similarly, in Newton's case, even if I grant you that he rejected certain physical theories, e.g., Cartesianism, in part because he thought them to be irreligious, would you say that religion was the motivation behind *all* his theory choices? For instance, was his decision to accept Boyle's law or his espousal of the theories of impact driven by similar religious concerns?

Relativist: I certainly wouldn't want to claim that Einstein's metaphysics or Newton's theology explains everything that each did as a natural scientist. But if you'll grant me that each of these beliefs shaped some of the central views of the scientist in question, that provides ammunition for my more general claim that elements of this sort enter crucially into virtually every important choice by scientists.

Positivist: But what I find so implausible about your general thesis, Quincy, is that there are plenty of scientific issues, perhaps most, which are entirely neutral with respect to large metaphysical or ideological issues. I mean, accepting the idea that gases expand when heated or the finitude of the speed of light or a certain value for Avogadro's number does not conjure up the same metaphysical or theological anxieties as the cases you were just describing.

Relativist: That's true, of course; but remember that among the elements which contribute to the social fixation of belief, ideology is only one. Even in much more mundane and ideologically neutral cases, scientists have matters of professional standing and prestige to consider—matters which may incline them to come down decisively one way or another on issues which you have been describing as "nonideological." The fact that an issue is remote from "high ideology" ought not suggest that a scientist has no self-interested stake in its outcome.

3. See Fine (1987).

Realist: There was a second preliminary matter I wanted to raise, and it applies to your use of both ideology and the professional interests of a scientist to explain his beliefs. Let us suppose for a moment that you were able to show that the theory preferences of a great scientist like Einstein were through and through driven by noncognitive concerns. If it was not his theology or philosophy, it might be his Judaism or his socialism or his keenness to win a Nobel Prize or his psychological makeup or what have you.[4]

Relativist: Keep it up, Karl, you'll turn yourself into a relativist yet!

Realist: The point I am making is that all of those factors at most concern themselves with a relatively uninteresting question: why did one man, Einstein in this case, come to hold a certain range of theories? Such issues are not what scientific epistemology is about. Our concern is, as Reichenbach used to put it, not with where theories come from but with what reasons, if any, we can give for holding them once they have been produced.[5] We're not interested in the psychopathology of the occasional scientist, however brilliant and influential he might be. Our concern is, or should be, with questions about why the scientific community as a whole came to accept these theories insofar as it did. Ideas have the craziest origins and those origins do nothing per se to indict the ideas in question. What you must show, Quincy, is that the major theories in science have been accepted by the *bulk* of scientists for reasons that had nothing to do with evidence, arguments, observations, and all those other goodies. I don't believe you—or your fellow relativists—have even begun to deliver on that task.

Relativist: No we haven't, at least not directly. But bear in mind

4. Compare Thomas Kuhn's remark: "Individual scientists embrace a new paradigm for all sorts of reasons, and usually for several at once. Some of these reasons—for example, the sun worship that helped make Kepler a Copernican—lie outside the apparent sphere of science entirely. Others must depend upon idiosyncrasies of autobiography and personality. Even the nationality or prior reputation of the innovator and his teachers can sometimes play a role" (1970, pp. 152–53).

5. For the locus classicus of this distinction, see Reichenbach (1938, chap. 1).

that what you're asking of us is a task that would require generations of research on these topics. What I am offering you, however, is a plausible argument from analogy. I'm saying that if it turns out—as I believe it does—that the theory choices of great scientists can be explained primarily in terms of their noncognitive interests, then it would not be the least bit surprising if the theory choices of lesser men were attributable to similar sorts of causes.

Realist: I see the analogy, of course; but I find it entirely unpersuasive for reasons that you have not grappled with. If you're right that a number of very specific facts about Einstein's social and political background explain why *he* adopted the theories he did—and maybe you're right about that—I cannot see how those same causes could possibly be invoked to explain why huge numbers of theoretical physicists, with backgrounds and interests vastly different from Einstein's, have also come to accept most of his theories. Englishmen and Americans, Christians and atheists, capitalists and libertarians, scientists from working-class and aristocratic families, young and old . . . the backgrounds of current-day advocates of, say, relativity theory cut directly across all the usual ways of categorizing noncognitive concerns. . . .

Relativist: You're on the verge of concluding that these other scientists couldn't have been "caused" to accept Einstein's theories by the same factors that made them look attractive to Einstein himself.

Realist: Precisely.

Relativist: Of course not. But it's a familiar philosophical principle that different constellations of causes can produce similar effects. The fact that a Lutheran miner in the Saarland and a Buddhist rice farmer in Vietnam have very different backgrounds does not prevent us from explaining, in class terms, why both become Marxists.

Realist: You're missing the moral, Quincy. The challenge that your approach faces is not that different causes can produce the same effects, it is rather this: each natural science is—at least much of the time—a highly *consensual* activity; at any given time, there is enormous agreement about many of the central theoretical commitments of the science in question. All of us except you explain that extraordinary degree of consensus by saying that scientists live in the same natural world, and that they generally abide by very similar rules for shaping their theory choices. By contrast, you have claimed, earlier in these conversations, that the world itself does virtually nothing to shape scientists' beliefs, and that the rules shared by scientists are so amorphous as to permit virtually any theory choices. That is why you argue for the necessity of invoking what you call "social causes" to explain scientists' beliefs.

Relativist: I'm still missing the point.

Realist: That's because I haven't stated it yet. It boils down to this: if different scientists have radically different personal agendas and interests, as you claim they do, and if those agendas rather than shared cognitive rules are the determinants of belief in the scientific community, which you also claim, then *it becomes utterly mysterious how strong consensus can arise in the scientific community.*

Relativist: Strong consensus arises in lots of communities, scientific and nonscientific; just consider Catholic monasticism.

Realist: But even in highly orthodox religions, there are rules governing which beliefs are permissible and which are not. I claim that it is the rules of method in science which play the same belief-determining role that authoritarian edicts do in highly regimented religions—with the crucial addendum that the methods of science are epistemically well founded whereas authoritarian epistemology is highly suspect.

Positivist: It is worth adding that consensus in science is the more remarkable because it is a discipline in which opinions are constantly changing—in sharp contrast to most religions, which rarely if ever admit of deep doctrinal changes.

Pragmatist: I'm not sure that the present line of exchanges can take us much further than it already has. So if I may, I'd like to pick up on a point which Quincy has referred to repeatedly: the question of what a scientist's "professional but extracognitive interests" are. Let us suppose for a second that scientists have no interest in finding true theories per se, or even empirically adequate theories per se. Let's suppose, moreover, that they are chiefly preoccupied with their standing and influence among their peers. Even in Quincy's tawdriest scenario, we can imagine how a scientist's professional self-interests might lead him to behave as if he were entirely objective and disinterested.

Relativist: You speak in paradoxes, Percy. Let's have the rest of it.

Pragmatist: Science appears to me to be a social activity in which the reward structure is heavily skewed in the direction of the researcher who picks theories which are *highly fertile*. It is those scientists who get the big grants, the waiting list of postdocs queued up to come and do research, the prestigious appointments. It is thus overwhelmingly in the *professional* interests of an ambitious researcher to make sure that the theories associated with him are perceived to be in the "fertile" category. You said earlier, Quincy, that it will always be in the professional interests of a researcher to continue to promote those theories in which he has already invested time and energy and with which his name may even be associated. It is thus impossible for you to explain, as it was impossible for Kuhn to explain, why a senior scientist should ever change his theoretical orientations. Indeed, Kuhn thought that one had to wait for the old generation to die off. But we now have ample evidence[6] that senior scientists often change their theoretical orientations *and* we have an explanation for why they do so; specifically, when they reckon that the theory they have been backing all these years has run out of heuristic steam, compared to its rivals. Where you say that it is in a scientist's professional interests to hang onto his present theory come hell or high water, I reply by saying that a scientist is heavily penalized professionally for proceeding in that fashion.

Relativist: So you're trying to tell me that the reward structure in science is so designed that entirely self-interested choices

6. See notes 9 and 10 below.

turn out to be the choices which promote the cognitive ends of the scientific enterprise?

Pragmatist: You've got it exactly. And this is what sensible sociologists of science—who have been largely ignored by you relativists—have been trying to tell us for more than a generation.

Relativist: And who designed science so that it would have this marvelous teleological feature?

Pragmatist: I don't know; maybe it just grew to be like that. But you really shouldn't be so cynical. We have loads of other social institutions where the reward structure encourages participants to perform in ways which conduce to the ends of the institution. Consider an institution like sport. We make it in an athlete's best interests to run faster, jump higher, and throw farther than he or she might otherwise be inclined to do just "for the love of it." Similarly in science, we have set up rewards and punishments which have the practical effect of keeping scientists more or less on the cognitive straight and narrow.

Relativist: I don't think I want to quarrel with your claim that scientists regard it as in their interests to be perceived as involved with programs of research which are flourishing and fertile. But you see that concern as explaining why scientists leap from one approach to another, why—in your language—scientists change their theoretical orientations. I peer at things through the other end of the telescope. I see any theory whatever as having the resources to make itself potentially fertile if enough bright people work on it. I don't want to drag us all through the holism business again but I do want to reiterate the point that we're never warrantedly in a position to judge that a theory has run out of heuristic potential, and I don't think any of you can seriously dispute that.

Positivist: I can. Where is the remaining heuristic potential in Aristotelian physics? For several generations after Galileo, natural philosophers desperately sought to revive a version of Aristotle's physics which could rival the achievements of seventeenth- and eighteenth-century mechanics. If repeated efforts by very bright and clever people have failed to grow flowers among the weeds, a time comes when we're entitled to infer that the soil is not very fertile.

Pragmatist: It's also important to stress that scientists, in general, are a very impatient lot. If an approach is bogged down, and failing to produce interesting new results, they quickly vote with their feet. That, I submit, is why theory change occurs so rapidly in the natural sciences.

Relativist: What you're alluding to is that scientists often decide, perhaps prematurely, that a certain approach has ceased to be fertile; but that's very different from proving that it is sterile.

Realist: Now who's talking about science as if it were a realm of disembodied ideas? I thought we were here today to try to understand why scientists adopt the beliefs they do. Percy has made the plausible claim that a practical way of formulating that problem is by asking why scientists change their theoretical orientations. He suggested that scientists abandon a theory when it is no longer producing interesting new results, that indeed it is in their professional self-interest to do just that. I find it ironic that you, Quincy, of all people are still trying to squeeze blood from the *logician's* stone, telling us that logic shows that any theory can be revived if only enough talented people will work on it. How, may I ask, does that point—even if correct—further your project of explaining why scientists adopt the beliefs they do?

Relativist: It's a well-known fact that scientists—especially older and better-established ones—tend to resist challenges to the prevailing paradigm (with which their careers have typically been bound up), ignoring its apparent anomalies and sticking with it through thick and thin. I'm not saying that they do this invariably; only that they do so with greater frequency than their younger colleagues. Such behavior looks irrational on a traditional view of what scientific methodology is. The argument from underdetermination, by blunting the impact of apparent refutations, shows why such behavior is rationally permissible.

Positivist: But by your account, Quincy, *virtually any behavior is permissible*. What you regard as the strength of your position, the fact that it countenances virtually any response to anomalies—from toughing it out by espousal of the old theory to adopting something completely new and untried—exhibits the poverty of your position. If anything goes, then nothing is explained.

Relativist: I haven't said that "anything goes." I have claimed that the evidence from the natural world does little if anything to explain scientists' beliefs about that world; that is because evidence does little if anything to constrain the scientists' theoretical orientations. But this does not lead to doxastic anarchy, for the central point that we relativists are making is that there are other forces working on a scientist besides the evidence and the rules of scientific method. It is these other causes which take up the slack left by the evidence in shaping scientists' beliefs. That's the whole point of talking, as we have been this afternoon, about social factors and interests and their key role in explaining the doxastic life of the scientist.

Pragmatist: And one example of such an explanation is your (and Kuhn's) hypothesis that older scientists are more resistant to challenges to the prevailing paradigm than younger ones because more of their prestige is bound up in maintaining the status quo?[7]

Relativist: Precisely, although I think the hypothesis is due originally to Max Planck.[8]

Pragmatist: If you were aware of the empirical studies that have been done in the last two decades on the so-called

7. Kuhn writes: "The transfer of allegiance from paradigm to paradigm is a conversion experience that cannot be forced. Lifelong resistance, *particularly from those whose productive careers have committed them to an older tradition of normal science,* is not a violation of scientific standards but an index to the nature of scientific research itself" (1970, p. 151).

8. Max Planck: "a new scientific truth does not triumph by convincing its opponents and making them see the light, but rather because its opponents eventually die, and a new generation grows up that is familiar with it" (1949, pp. 33–34).

"Planck principle," you'd be a little less confident about your hypothesis. Although some of the studies seem to bear out the Planck hypothesis, an equal number do not reveal any striking positive correlation between a scientist's age (or tenure in the profession) and his resistance to radical new approaches.[9] In one such study, for instance, it turned out that the early converts to Darwin's evolutionary theory were actually *older* than its detractors.[10]

Relativist: That may well be so; I'm certainly not wedded to the age-conservatism hypothesis. I cited it only as a well-known example of how extra-evidential, i.e., social, hypotheses have got to be brought in to explain what evidence itself cannot—the vicissitudes of scientific belief. If you have any doubts about the general fertility of this approach, you should look at the hundreds of studies that have been done in the last two decades, documenting the role of class, ideology, nationality, prestige, and professional self-interest in shaping the attitudes of natural scientists. I refer you particularly to a summary of many of those studies recently written by Steve Shapin.[11]

Positivist: I want to see if I'm getting this right. You're trying to tell us now that there is a lot of empirical research to "support" the claim that evidence plays almost no role in shaping scientists' beliefs?

Relativist: Exactly!

Positivist: And you're trying to tell us that we should accept this hypothesis of yours because there is such impressive evidence for it?

Relativist: Quite.

Positivist: But, Quincy, if you were true to your avowed convictions, you would eschew any attempt to find evidence for anything, since you believe evidence to be irrelevant to fixing beliefs. The fact is that you will use empirical evidence when you find it rhetorically convenient, even if it is in aid of the

9. See, for instance, James Hofmann (1988), and Arthur Diamond (1988).
10. David Hull, Peter Tessner, and Arthur Diamond (1978).
11. Shapin (1982).

claim that evidence never makes much difference. You relativists can't get your act together on this issue, can you? You like to present yourselves in the guise of empiricists and naturalists, simply reporting on what the facts are. But one of the key "facts" that you attempt to establish is that facts are open to more or less any interpretation. Surely you can see—to put it in the most charitable possible fashion—that your methods of research and your research conclusions scarcely sit well together.

Pragmatist: I think, Rudy, that the point comes up vividly in connection with the social-interests model which Quincy has been describing for us. The interest model tells us, if I understand it rightly, that scientists make their choices of theory so as to serve their own social and professional interests. Now, if that is so, I suppose that we are to imagine a scientist engaged in a process of mental calculation in which he contemplates various courses of action open to him—accept this theory, reject that one, etc.—and then selects that action which best promotes his interest. Have I got that right, Quincy?

Relativist: Yes, as far as it goes.

Pragmatist: Well, my question for you is this: how does the scientist decide which action is in his best interest? It couldn't be by appeal to any inductive or empirical information about which means in the past have furthered his ends since—on your view—the relevant evidence would always support more or less any hypothesis about which means promote which ends. So what else is available to the scientist for him to decide what promotes his social interests?

Relativist: I'm not sure how scientists decide what is in their self-interest. I'm not even sure that the calculation is a conscious one. But what I am sure of is that scientists generally act in what they regard to be their professional self-interest.

Pragmatist: Well, let's leave the scientist out of it for a moment. How do *you* as a relativist—concerned to explain scientists' behavior in terms of their interests—how do you figure out what a scientist's self-interests are?

Relativist: I don't need to know what a scientist's genuine self-interests are to get my explanatory equipment moving; all I need to know is what a scientist regards as being in his self-interest.

Pragmatist: Fine, but how do you find *that* out? To say of a scientist that "Mr. X believed his self-interest would be promoted by Y," I would have thought that you needed some evidence about Mr. X's beliefs, intentions, status in the scientific community, prior training, and so on.

Relativist: Yes, of course I do; what point are you making?

Pragmatist: Simply this: your explanatory project cannot get off the ground without propounding *hypotheses* about scientists' interests, beliefs, careers, and so on. Yet the principle of underdetermination of which you are so fond entails that no one of those hypotheses attributing interests to scientists is any better established than any rival attributions of interest might be. Indeed, it says that—compatibly with *everything* we might come to know about a scientist—*any* hypothesis attributing beliefs, intentions, or interests to a scientist is as good as any other. That means, among other things, that we might suppose a scientist's aesthetic interests or his cognitive interests or anything else to be the causes of his actions. And that in turn means that, by your own lights, the so-called model of social interests can never be said to be better established than theories which deny that scientists ever act so as to promote their social interests.

Relativist: I'm afraid that I just don't see the bearing of all these meta-level worries on a project of empirical research into the social roots of scientific belief which has been extraordinarily successful during the last two decades.

Realist: Maybe I can come at it from a slightly different direction. You just referred, Quincy, to this extensive body of empirical research which bears out your claims. This research you refer to, Quincy, who is it that has been doing it?

Relativist: Primarily sociologists and social historians of science, along with the occasional anthropologist or philosopher.

Realist: These are all scholars who presumably have a professional self-interest in establishing the idea that social causes are much more influential than cognitive causes in determining scientific belief. Would that be fair to say?

Relativist: That question is purely ad hominem.

Realist: Of course it is; but your chickens are coming home to roost, Quincy. The whole thrust of your arguments over the last several days has been that rational deliberations about evidence are always inconclusive. You tell us further that the real precipitating cause for a scientist to accept or reject a theory or paradigm is his calculation as to whether it will serve his professional and extraprofessional interests. By your own lights, therefore, if I want to explain why *you* hold the beliefs you do, I have to inquire about how those beliefs further your position and ambitions, the standing of your profession, the interests of the groups of which you are a part. You cannot cry foul when I suggest that you endorse the relativist position, not because of evidence and arguments, but because it promotes your parochial interests. That, after all, is the only sort of story about beliefs which your principles will countenance.

Pragmatist: I'd like to come at it from a rather different angle which relativists like Quincy may find slightly more congenial. I mean the political angle. I hope that I hardly need remind you that relativism has been taken up enthusiastically by a variety of political and quasi-political movements?

Realist: Perhaps you have in mind certain feminist groups that are concerned to show that science, scientific rationality, the experimental method, and a host of other features associated with science merely reflect the interests of the men who have wrought the scientific picture?

Positivist: You might add to your list "the cultural left," who generally think that Western culture and its chief artifacts, science and technology, represent a retrograde step.

Pragmatist: The list could be made significantly longer if we thought about it for a while. But I was wanting to make a point here, and the examples you've already mentioned will be ample. The point is that these groups are undermining their position by adopting strongly relativist stances.

Relativist: How can you say that? It is the relativist perspective that has made, say, the feminist critique of science viable. So long as scientists could persuade the rest of us that they were simply reporting how the world was, it became impossible to motivate the claim that gender-based issues were warping the scientific perspective. Now that relativists have shown that there is much more to science than meets the eye—pun intended—it becomes possible to explore the sources of these biases.

Pragmatist: Quite the contrary, Quincy. To say that a certain perspective is biased—and surely we do want to say that from time to time—is to claim that it deviates from an objective report about how things are. To justify a charge of bias requires us to be able to compare the allegedly biased report with an unbiased one. But if all reports are equally defensible—if every claim merely reflects the interests of its advocates—then claims of bias cease to have any pejorative force.

Realist: There is another permutation here. Charges of bias must themselves be appraised. If one wants to claim that the experimental method has a male bias to it, one must be able to marshall arguments and evidence to support such a charge. But if evidence is open to any interpretation one likes, then evidence and argument likewise cease to have any genuine role in politico-cultural debate. Think of it this way: feminists and other cultural critics, whether on the left or the right, want to make some positive claims—claims about matters of fact. Feminists, for instance, want to argue that Western science exhibits some strong gender-based perspectives.

Relativist: And so?

Pragmatist: Such claims carry little force unless they are grounded in a careful and scholarly study of the record. Karl Marx and several generations of scholars on the political left taught all of us that lesson, right?

Relativist: Of course, but I still fail to see the point.

Pragmatist: It's just this: if everything one says is just a matter of one's perspective, if any point of view is as good as any other, then one is never going to be in a position to argue that any political agenda is preferable to another. If you relativists are right, then we are forced to say, for instance, that the hypothesis about the gender bias of science is just one hypothesis, neither better nor worse founded than the claim that science has no gender bias at all.

Positivist: And the same point applies, of course, to all the other items on the agenda of the new left. Claims like "corporations pollute," "nuclear power is unsafe," "the environment is deteriorating," "centralized authority is dangerous," "there is systematic discrimination against women, blacks, or other minorities"—none of these claims can carry any authority if all beliefs are on a par.

Relativist: You seem to be supposing that slogans of this sort carry force only if one can show them to be true. You ignore the fact that most of political life is rhetoric and persuasion; it's a matter of numbers, not arguments.

Positivist: I don't ignore that in the slightest. But those on the left should be especially wary of that way of proceeding. The left has generally found itself in the minority in Western culture. It has consoled itself with the hope that by informing people about how matters *really* stand—whether those matters have to do with class structure, racism, sexism, or what have you—it could win a larger following for its perspective.

Relativist: But the left still engages in such campaigns.

Positivist: Indeed; but to the extent—and it is considerable—that the new left subscribes to strong forms of relativism, it has lost all theoretical rationale for such activity. What I am saying is that if the radical feminists, counterculturists, and others were to acknowledge in the public arena their theoretical convictions, they would immediately lose any public following *and deservedly so*. I mean such convictions as that texts and statements have no determinate meaning, that no hypotheses about political economy or race relations or gender issues are any better supported than their denials, or that claims in the public arena ostensibly about matters of fact reflect nothing about the facts of the matter. The sad truth is that relativism can no more sustain a political agenda than it can underwrite a scientific one.

Pragmatist: I wonder if we couldn't turn to a rather more constructive side of this topic? The relativist, if I understand his position correctly, despairs of epistemology because it purports to tell us what we ought to believe rather than to describe how we come to believe what we do. Mainline epistemology, by contrast, has been neglecting to explain why people believe what they do.

Relativist: It goes further than that, Percy. If there is any legitimate role for the theory of knowledge in our era, it is one modeled on the sciences. As I said earlier, scientific knowledge is not about what we ought to do; it attempts to describe what is.

Realist: Such language, coming from you, Quincy, is pure hypocrisy. You fervently deny that we have any access to how things really are; so how can you say that science "describes what is"?

Relativist: I shall ignore that ad hominem. As I see it, the challenge facing us is to explain how systems of belief come to be developed, accepted, and rejected, providing a kind of natural history of belief. Beliefs, whether we like them or dislike them, believe them true or believe them false, should all be explained. The relativists' cri de coeur is that the project of explaining the anatomy of belief is far more important than presuming to tell scientists how they should reason about the world. You may not like the "interest models" that I have been alluding to for explaining beliefs, but I have yet to hear any argument against the legitimacy of the relativists' project for explaining the fortunes of beliefs, both scientific and nonscientific.

Pragmatist: I don't think any of us have quarrels with that project per se, although I myself might have imagined that such was more properly the task of the psychologist rather than the philosopher or the sociologist. What I think we all object to, Quincy, is the suggestion that such chores exhaust the legitimate functions of the theory of knowledge.

Relativist: That's surely because, as I explained this morning, you still believe that the epistemologist has access to special forms of authentication and justification which go beyond the conceptual scheme in which we find ourselves located. It reflects the fact that you three are interested in normative questions, not factual ones.

Pragmatist: Not so. Let me take you literally when you say that the primary function of the theory of knowledge is to explain features of systems of belief. Even on an explanationist agenda, there are crucial facts about certain systems of belief which you refuse to address.

Relativist: Such as?

Pragmatist: An issue we've touched on before—the empirical success of science. What impresses Rudy, Karl, and me about scientific belief is that it confers predictive and manipulative control on those who accept it. To note that fact is not necessarily to endorse it; I need not commit myself for these purposes as to whether such control over nature is a good thing or a bad one. It is a descriptive property of the natural sciences that they provide such control. Now we three think it's pretty important to explain that fact about scientific knowledge; indeed, we see it as the central task of scientific epistemology to tackle that issue. That doesn't mean we're agreed about the solution; Karl and I in particular fail to see eye-to-eye about how to explain the success of science. By contrast, in insisting that all systems of belief—religious and scientific, political or philosophical—are to be explained in the same way, you adopt an approach which insures in advance that you lack the resources to tackle the problem of explaining the success of science.

Relativist: I'm not at all sure that I agree that science has been especially successful, or, even if it is successful, that such success testifies to more than its cultural imperialism.[12] But leaving that to one side, I don't see why it follows that relativism lacks the resources in principle to explain such differential success as science exhibits.

Pragmatist: What I meant was this: so far as I can see, the explanation of the empirical success of science is going to have to

12. Feyerabend again: "Today science prevails not because of its comparative merits but because the show has been rigged in its favor" (1978, p. 102).

be sketched, at least in part, in terms of the linkages between us, our beliefs, and the natural world. If the world did nothing whatever to shape and inform our beliefs about it, it would be absolutely extraordinary if our theories managed to work as well as they do.

Realist: That sound like realism to me, Percy.

Pragmatist: It is realism of sorts, Karl, but manifestly not your sort. You seek to explain the success of science by supposing that scientific theories are true. "If," you argue, "our theories were true—or approximately true—they would be successful; otherwise, the success of science is a miracle."[13] I readily grant that if our theories were true, they would be successful; but since I think they're almost surely not true, I don't find your account very convincing, Karl, despite its appeal in the abstract. By contrast, I want to tell a story that goes something like this: we find ourselves in a situation where our only contact with the world is mediated by our concepts. We posit certain beliefs or theories to make sense of that mediated world. If those beliefs or theories were entirely free-floating (as Quincy believes them to be) and reflected nothing whatever about the world itself, then it would be unthinkable that they would enable us to manipulate the world as effectively as we can. For my money, the explanation of the success of science is going to have to be told in terms of the ways in which our interaction with nature puts strong constraints on our systems of belief.

Relativist: But even if you can fill in the details of that very sketchy story, Percy, it remains but one among indefinitely many hypotheses for explaining the success of science.

Pragmatist: It may be only one among several, but they're not

13. Hilary Putnam claims that unless our theories are approximately true, "the success of science is a miracle" (1978, p. 69). Ernan McMullin writes: "The claim of a realist ontology of science is that the only way of explaining why the models of science function so successfully . . . is that they approximate in some way the structure of the object" (1970, pp. 63–64). Newton-Smith similarly takes the line that the increasing predictive success of science would be completely mystifying if it were not for the fact that theories are capturing more and more truth about the world.

exactly thick on the ground. More to the point, I've yet to hear even the most tentative sketch of how someone with your philosophical orientation might tackle the problem.

Relativist: I've already conceded that I don't have a special account of the success of science but that concession arises, in part, from the fact—to which I've alluded many times—that I don't think that the success of science is either unique to science or in crying need of explanation. As Rorty has rightly noted, philosophers have long supposed that there is some secret to the success of science, when the fact is that there is none.[14]

Realist: When you say there is "none," do you mean that science is not notably successful or that it is successful but that there is no "secret" to its success?

Relativist: Probably a little of both. Insofar as science is successful, that success involves its ability to predict and manipulate natural objects and events. But as Rorty once asked, "what is so special about prediction and control?"[15] Moreover, there are many different notions of success. Consider how spectacularly successful Islam is at winning converts, or the progress of communism in African nations.[16] No one regards those successes as being in need of exotic philosophical story-telling. Why then should we regard the success of science any differently?

Positivist: What's called for, surely, is a distinction between success at persuading others to join one in a cause—which is what Quincy's examples speak to—and success at anticipating nature. Of course, we want explanations for both sorts of phenomena. But whereas it seems appropriate to explain the first type of success using, perhaps, principles of rhetoric and social psychology, neither of those disciplines can possibly explain why science is so successful at predicting natural events. In arguing that philosophy has a natural role to play in explaining the success of science, we are not denying that

14. Rorty (1980, p. 55).
15. Rorty (1988, p. 66).
16. These examples are Rorty's (1988, p. 61).

there are other forms of "success" or asserting that it is unimportant to explain them. But I'm damned if I can see how one can reduce the explanation of the empirical success of science to a question of applied social science.

Relativist: But whether a theory triumphs in the scientific community is just as much a matter of rhetoric and persuasion as whether Islam wins converts.

Positivist: I doubt it very much, but even if you're right that's wholly beside the point since the success we are attributing to science is not that of winning a loyal following among the community of scientists but that of anticipating nature's twists and turns. That sort of success has nothing whatever to do with persuasion. Surely you can see the difference without my having to draw you a picture. . . .

Relativist: Of course I can, but that takes me back to my earlier point. Even if we focus on prediction and control as a particular form of success, there is nothing very surprising about the fact that, after some two thousand years of trying, scientists are reasonably successful in that regard. As Feyerabend pointed out some years ago, it just stands to reason that if you have enough clever human beings engaged in any activity, it's bound to produce impressive results sooner or later.[17]

Positivist: Oh, really? And what are the "impressive results" of astrology, psychical research, or theology, each of which can count numerous "clever human beings" among its practitioners?

Relativist: It depends on whom you talk to, and what you regard as impressive or "successful." I daresay that more people read their horoscopes, act on what you'd call their superstitions, and say daily prayers than ever accepted the theory of evolution or quantum mechanics.

Positivist: Here comes your conflation of different forms of success again. I want to know what empirical success astrolo-

17. Feyerabend: "An empirical theory (as opposed to a philosophical theory) such as quantum mechanics . . . can of course point to numerous positive results, but note that *any view* and *any procedure* that is developed by intelligent human beings has results" (1981, pp. 140–41; my emphasis).

gy has exhibited at predicting the future, despite the fact that loads of very bright people, e.g., Kepler and Galileo, put their minds to doing so. Or where is the evidence that prayer enables us to anticipate or to manipulate natural events?

Relativist: Ask that question to millions of "born-again Christians" and they will tell you that the evidence is there in their daily lives. . . .

Pragmatist: Gentlemen, I have the strong impression that we are beginning to repeat ourselves. I've also got my eye on the clock since we all have flights to catch this evening. I wonder if, before we go our separate ways, we shouldn't discuss briefly how we will assemble the report for the American Philosophical Congress?

Realist: Quite so.

Pragmatist: I think that we've probably covered most of the relevant topics in these three days but I'm not sure that anything approaching a consensus has emerged. What do you think?

Positivist: I think it's altogether clear that relativist epistemology has little to recommend it. Its central tenets are often fuzzy and when they're not, they're wrongheaded.

Realist: I think I have to agree with Rudy. As he has pointed out over and over again, relativism is self-referentially incoherent, and to boot it is predicated on several dubious epistemological theses—for instance, incommensurability, holism, and radical underdetermination. The relativist supposes that fallibilism, which we all accept, entails that all beliefs are equally well- or ill-founded. Except his own, of course, which he believes to be better-founded than ours!

Pragmatist: Need I bother to ask, Quincy, whether you are a party to this emerging consensus?

Relativist: Of course I'm not. I grant that you have made some telling arguments against various tenets of relativism, but all philosophical positions—not least your own—face all sorts of anomalies and other difficulties. For instance, realism continues to work with a notion of some sort of mysterious "correspondence" between beliefs and the world which Karl has done nothing to clarify; nor has he shown that realism is very well situated to explain much of interest about the scientific enterprise. Rudy's self-styled opposition to fuzziness sounds fine when applied to the views of others, but he rarely brings it to bear on his own pet projects. For instance, he still operates with the notion of a clear distinction between the "observational" and the "theoretical," when we all know that dualism will not stand up to scrutiny. Similarly, he and his fellow positivists continue to suppose, despite all their disclaimers, that the empirical base can legitimately be regarded as a veridical foundation for knowledge. As for you, Percy, your tightrope act, which attempts to balance between realism and relativism, is an inherently unstable position. To change my metaphor, you're wanting to have your cake and eat it and you simply cannot do both. Besides, to return to my own position for a moment, I have heard nothing in these three days that persuades me that relativism is, as some of you seem to think, rotten to the core.

Pragmatist: Returning to the immediate business at hand, is it agreeable that I should write a draft of the committee report and circulate it to the three of you for comments and revisions? If necessary, Quincy can write his own minority report if he continues to find himself at odds with the rest of us. . . .

References

Aiton, E. 1972. *The vortex theory of planetary motions.* London.

Barnes, Barry. 1982. *T. S. Kuhn and social science.* New York: Columbia University Press.

Bloor, David. 1976. *Knowledge and social imagery.* London: Routledge.

———. 1982. "Reply to Buchdahl." *Studies in history and philosophy of science,* 13:305–11.

Boyd, Richard. 1973. "Reason, underdetermination and a causal theory of evidence." *Nous,* 7:1–12.

Carnap, Rudolf. 1931. "Überwindung der Metaphysik durch logische Analyse der Sprache." *Erkenntnis,* 2:219–41.

Davidson, Donald. 1984. "On the very idea of a conceptual scheme." In Davidson, *Truth and Interpretation.* Oxford: Oxford University Press.

Diamond, Arthur. 1988. "The polywater episode and the appraisal of theories." In A. Donovan et al., pp. 181–98.

Donovan, A., et al. 1988. *Scrutinizing science.* Dordrecht: Kluwer.

Doppelt, Gerald. 1978. "Kuhn's epistemological relativism." *Inquiry,* 21:33–86.

Feyerabend, Paul. 1975. *Against method.* London: New Left Books.

———. 1978. *Science in a free society.* London: New Left Books.

———. 1981. *Realism, rationalism and scientific method.* Cambridge: Cambridge University Press.

Fine, Arthur. 1987. *The shaky game.* Chicago: University of Chicago Press.

Goodman, Nelson. 1955. *Fact, fiction and forecast.* Cambridge: Harvard University Press.

Grünbaum, Adolf, and W. Salmon, eds. 1988. *The limitations of deductivism*. Berkeley: University of California Press.

Harding, Sandra, ed. 1976. *Can theories be refuted?* Dordrecht: Reidel.

Hempel, Carl. 1965. *Aspects of scientific explanation*. New York: Macmillan.

Hofmann, James. 1988. "Ampère's electrodynamics." In A. Donovan et al., pp. 201–17.

Hull, David, Peter Tessner, and Arthur Diamond. 1978. "Planck's principle." *Science*, 202:717–23.

Kuhn, Thomas. 1970. *Structure of scientific revolutions*. 2d ed. Chicago: University of Chicago Press.

———. 1977. *The essential tension*. Chicago: University of Chicago Press.

Lakatos, Imre. 1978. *The methodology of scientific research programmes*. Cambridge: Cambridge University Press.

Lakatos, Imre, and A. Musgrave, eds. 1970. *Criticism and the growth of knowledge*. Cambridge: Cambridge University Press.

Laudan, Larry. 1981. *Science and hypothesis*. Dordrecht: Reidel.

———. 1984. *Science and values*. Berkeley: University of California Press.

Lukes, S., ed. 1982. *Rationality and relativism*. Cambridge: MIT Press.

McMullin, Ernan. 1970. "The history and philosophy of science." In R. Steuwer, ed., *Minnesota studies in philosophy of science*. Minneapolis: University of Minnesota Press.

Nicholas, John. 1988. "Planck's quantum crisis and shifts in guiding assumptions." In A. Donovan et al., pp. 317–36.

Planck, M. 1949. *Scientific autobiography and other papers*. New York: Philosophical Library.

Popper, Karl. 1959. *Logic of scientific discovery*. London: Hutchinson.

Post, Heinz. 1971. "Correspondence, invariance and heuristics." *Studies in history and philosophy of science*, 2:213–55.

Putnam, Hilary. 1978. *Meaning and the moral sciences*. London: Routledge.

Quine, W. 1969. *Ontological relativity and other essays*. New York: Columbia.

———. 1970. "Grades of theoreticity." In Swanson and Foster, eds., *Experience and theory*. London: Duckworth, 1970.

———. 1976. "Two dogmas of empiriciam." In S. Harding, ed., pp. 41–64.

———. 1981. *Theories and things*. Cambridge: Harvard University Press.

Reichenbach, Hans. 1938. *Experience and prediction*. Chicago: University of Chicago Press.

Rorty, Richard. 1979. *Philosophy and the mirror of nature*. Princeton: Princeton University Press.

———. 1980. "A Reply to Dreyfus and Taylor." *Review of metaphysics*, 34:39–46.

———. 1988. "Is science a natural kind?" In E. McMullin, ed., *Construction and constraint*. Notre Dame: University of Notre Dame Press, pp. 49–74.

Shapin, S. 1982. "History of science and its sociological reconstructions." *History of science*, 20:157–211.

Winch, Peter. 1964. "Understanding a primitive society." *American philosophical quarterly*, 1:307–24.

———. 1970. "Comment." In R. Borger and F. Cioffi, eds., *Explanation in the behavioral sciences*. Cambridge: Cambridge University Press.

Index

Aims of science, 18, 19, 27, 52, 53, 100
Aiton, E., 113n
Ampliative inference, 67, 74, 99
Anomalies, 71, 82, 89, 156, 157
Aristotle, 83–85, 129, 155
Articulation of a paradigm, 72
Auxiliary assumptions, 57, 73, 76–84
Azande, 116–19

Barnes, Barry, 55n, 106n
Belief: social determination of, 146–66; stability of systems of, 55n, 115; standards of evaluation of, 106–12, 114, 136
Bias, 162, 163
Bloor, David, 55n, 74, 106n, 135, 148n
Bohr, Niels, 9
Boyd, Richard, 17

Carnap, Rudolf, 58, 88, 95, 97
Cartesian paradigm, 90, 91, 113
Causation, 10, 117, 149, 150
Charity, principle of, 129, 130
Classical mechanics (Newtonian mechanics): and Galileo's and Kepler's laws, 21–23; incommensurability with relativity theory, 124, 125, 129, 130; as limiting-case of relativity theory, 9–12; and vortex theory, 16, 25; replacement by relativity theory, 89, 113
Collins, Harry, 74
Commensurability of theories, 139, 140. *See also* Incommensurability of theories
Compatibility of theories with observations, 48–51, 80
Conceptual change. *See* Paradigm change
Conceptual scheme, 90, 125, 126, 129, 165. *See also* Paradigm
Confirmation: and fallibilism, 57; and holism, 71, 79, 80; of methodological rules, 104; and scientific progress, 6, 13; and theory testing, 21, 61, 63, 64
Confirming instances, 21, 23, 61–63, 65, 67
Consensus, 152, 153
Consistency, 136
Convention: and falsification of theories, 87, 88; and observation reports, 44, 45; and rules of science, 66, 97, 99–101, 103; and standards of theory choice, 57
Copernicus, Nicolaus, 90
Corrigibilism. *See* Fallibilism
Crucial experiment, 143
Cultural imperialism, 109, 165
Cultural relativism, 115
Culture, 101, 108–19, 148n
Cumulativity, 14–20, 27, 31. *See also* Explanatory loss

Darwin, Charles, 90
Davidson, Donald, 125, 126n

Deductive logic: and nondeducibility of theories from observations, 49, 56–58, 104; and scientific method, 99; and theory choice, 74. *See also* Entailment
Deductive-nomological model, 51
Deep structure theories, 37, 41
Descartes, René: vortex theory of, 15, 16, 25. *See also* Cartesian paradigm
Dewey, John, 134
Diamond, Arthur, 159n
Doppelt, Gerald, 75n, 76n
Duhem, Pierre, 5, 35, 56–58, 87
Duhem-Quine thesis, 41, 42, 76, 77, 83n

Economy, principle of (Ockham's razor), 137, 138
Einstein, Albert, 9, 149–52
Empirical adequacy of a theory, 60, 61, 64, 144, 145
Empirical consequences of a theory (observational consequences): comparison of theories by, 3, 4, 7, 9, 10, 14, 16, 22, 140; of holistic systems, 81; theory testing by, 24, 31, 61, 64, 65; and underdetermination, 49, 50, 52
Empirical equivalence of theories, 65
Empirical research: on justification of scientific method, 96, 158, 160
Empirical success of science: as problem of epistemology, 165–68; and scientific method, 94, 116–19, 137, 144; and scientific progress, 3, 13, 15–17
Empiricism: and observation reports, 44, 46, 47; and relativism, 159; and theory-ladenness, 35, 36
Ends and means, 104, 114
Entailment: and comparison of theories, 4–7, 9, 10, 15, 16, 19–21; and explanation, 52; and raven paradox, 63; and underdetermination, 50
Epistemic equivalence of theories, 65
Epistemic relativism, viii, 115
Epistemology: as descriptive social science, 46; and foundationalism, 86; and incommensurability, 133–38; progress of science accounted for by, 2; rational concerns of, 151; and standards of methodological rules, 103, 105–7; success of science accounted for by, 164, 165; and theory testing, 33
Ethnocentrism, 101
Evaluation. *See* Valuation
Evidence: and holistic systems, 74, 78, 81, 86, 87, 89; and interests, 147, 148, 151, 158, 159, 161; and raven paradox, 63; standards of, 56; theory-ladenness of, 34–36, 38, 43, 44, 46; and theory testing, 22, 23, 29, 105, 112; and underdetermination, 50, 51, 54, 55
Evidence report. *See* Observation report
Experiment: and confirming and positive instances, 63; and holism, 70; mistakes in, 136; and observational and target theories, 47; and theory-ladenness, 38; theory testing by, 19. *See also* Crucial experiment; Tests of a theory
Experimental laws, 7–10, 43
Explanation: as aim of science, 18; and auxiliary assumptions, 79–82; and entailment, 52; and holistic systems, 86; and scientific progress, 3, 14; of success of science, 164, 166, 167; and theory testing, 19–23, 31; and underdetermination, 51
Explanatory loss, 15–17, 24–27, 33

Fallibilism (corrigibilism), 38, 57, 132, 133, 169
Falsifiability, 5, 55n

Falsification: and comparison of theories, 14, 140, 141; and convention, 45; and holism, 71, 72, 87–89; and nondeducibility of theories from observation, 57, 58; and theory-ladenness, 39–42
Feminism, 161–63
Fertility of theories, 154, 155
Feyerabend, Paul: empirical success of science, 94, 168n; evidential equivalence of theories, 55; explanatory loss in theory change, 15, 25; incommensurability, 121, 122, 124, 140; noncognitive account of science, 165n; observational and target theories, 47; paradigm shifts, 89, 90, 92; theory and worldview, 12
Fine, Arthur, 149n, 150n
Foundationalism, 45, 86, 100, 108, 135

Galileo Galilei, 90
Galileo's laws on free fall, 21–23
Generality of a theory, 4–7
Gestalt, 73, 89n
Goodman, Nelson, 63, 64
Grue paradox, 63, 64
Grünbaum, Adolf, xi, 52n, 77n

Hempel, Carl, 22, 51, 58, 63
Historical induction. *See* Pessimistic induction
Hofmann, James, 158n
Holism, 57, 69–81, 86, 89–93, 111
Hull, David, 158n
Hume, David, 56–58, 104
Huygens, Christian, 91
Hypothesis. *See* Theory
Hypothetico-deductive method, 99

Ideology, 106, 150, 151
Imperatives, 98, 102
Implicit theory of meaning, 70, 131
Incommensurability of theories (indeterminacy of translation), 12, 121–45
Incorrigible given (indubitable given), 134, 135
Indeterminacy of translation. *See* Incommensurability of theories
Indubitable given. *See* Incorrigible given
Inductive logic, 58, 59, 67, 99. *See also* Ampliative inference
Infallibilism, 86. *See also* Fallibilism
Instrumentalism, 31, 36, 37
Interests, social. *See* Social interests

Justification, 95, 105–107, 134, 136, 137, 164
Justificatory ascent, 102

Kant, Immanuel, 16
Kepler, Johannes, 151n
Kepler's laws of planetary motion, 21–23
Kitcher, Philip, xi
Kuhn, Thomas S.: evidential equivalence of theories, 55; explanatory loss in theory change, 6n, 15, 17, 25, 27; holism, 71–73, 76–78; incommensurability, 121, 122, 124, 140; paradigm shifts, 89, 90, 92, 154, 157; relativism disavowed by, xi; social interests in science, 151n; standards of paradigm selection, 98, 112, 113n, 116; theory and worldview, 12

Lakatos, Imre, 54, 55n, 73, 83, 87, 89, 90
Language: and incommensurability, 124, 125, 127–30; of observation reports, 45; and theory-ladenness, 35. *See also* Meaning
Laplace, Pierre-Simon, 16
Laudan, Larry, 17n, 39n, 90n

Left, the, 161–63
Limiting-case relations, 9–11, 13, 15, 17
Loss. *See* Explanatory loss

Maxwell, Grover, 35
Mayo, Deborah, xi
McMullin, Ernan, 166n
Meaning, 43, 70, 124
Methodological rules: conventional nature of, 88, 101; epistemological status of, 94–112; of paradigms, 77, 78; and theory testing, 34, 35, 59, 93; and underdetermination, 66, 67. *See also* Scientific method

Naturalism, 134, 135, 159
Naturalistic fallacy, 98, 99
Nebular hypothesis, 16
Neurath, Otto, 35
Newton, Isaac, 11, 12, 90, 91, 149, 150
Newton-Smith, W. H., 166n
Newtonian mechanics. *See* Classical mechanics
Nicod criterion, 22–24, 63, 74
Nicholas, John, 113
Normal science, 116
Normative epistemology, 135, 137, 138, 164

Objectivity, 93, 94, 98, 142, 149, 154, 162
Observation: incommensurability of terms of, 124; as raw material of science, 3; and theory, 20, 22, 43, 63, 135, 151, 170; theory-ladenness of, 8, 34–38. *See also* Observation report
Observation language, 128
Observation report (evidence report), 36–38, 40, 41, 44, 45, 48
Observational consequences. *See* Empirical consequences of a theory
Observational theory, 47
Occasion sentence, 132
Ockham's razor. *See* Economy, principle of
Ontology, 73, 74, 125. *See also* Theoretical entities

Paradigm: and evidential equivalence of theories, 54; and holism, 71–92; incommensurability of, 121–26, 128–32, 139; and problem solving, 27n; and social determination of belief, 161; standards for evaluation of, 109–13, 115, 116, 140–42; and versions, 72. *See also* Conceptual scheme
Paradigm change (conceptual change), 115, 116, 122n, 151n, 156, 157
Paradoxes of induction, 58, 67
Partial incommensurability, 122–24, 131, 139–41
Peirce, Charles Sanders, 2, 88
Persuasion, 163, 167, 168
Pessimistic induction (historical induction), 39
Phlogistic chemistry, 89
Physics, 15. *See also* Classical mechanics; Relativity theory
Planck, Max, 9, 113, 157, 158
Poison oracle, 116–19
Politics, 161–63
Popper, Karl: acceptance of theories, 59; conventionality of rules, 97, 100; falsifiability, 87, 88; inductive logic, 58; normative epistemology, 135; progress of science, 2; theory-ladenness, 35, 44
Positive instances, 24, 61–63, 65–67
Positivism: decline in philosophy of science of, vii–viii; scientific progress viewed by, 2–3, 13; and subjectivity of methodological rules, 100; and theory-ladenness, 35, 36, 170; and underdetermination, 53
Post, Heinz, 14n

Pragmatism: criteria for theories of, 28, 101; and normative epistemology, 134; and underdetermination, 53. *See also* Instrumentalism
Prediction: as aim of science, 18, 103; and holistic systems, 71–73, 77–80; and paradigm shifts, 88, 89, 117–19; and success of science, 165, 167, 168; and scientific progress, 3, 6, 7, 26. *See also* Surprising predictions
Probability, 86
Problem solving, 27–30
Professional interests, 148, 150, 151, 153, 154, 156, 161
Progress of science, 2–32, 53, 89
Pseudo-science, 107
Putnam, Hilary, 17, 39n, 49n, 59, 101, 166n

Quantum theory, 9, 113
Quine, Willard van Orman: empirical adequacy, 60, 61; falsifiability of theories, 5, 42, 71, 72; holism, 76, 77; incommensurability, 121, 124, 134; relativism disavowed by, xi; scientific method, 136n; theory-ladenness, 44; underdetermination of theories by observation, 65. *See also* Duhem-Quine thesis

Rationality: in accepting scientific theories, 16, 86, 87, 140, 156, 161; and change of holistic theories, 74, 76n, 83, 84; and incommensurability, 128, 139; and methodological rules, 93, 94, 114, 115, 119. *See also* Reasons
Raven paradox, 63, 64
Realism: and empirical equivalence of theories, 65; and falsification, 87; and normative epistemology, 135; and observation and theory, 8, 35, 37; scientific theory viewed by, 12, 13, 166, 169; and standards of theory evaluation, 59; and subjectivity of methodological rules, 100; and underdetermination, 49, 52, 53
Reasons, 16, 50, 65, 76, 147, 151
Refutation of a theory: and holistic systems, 72, 78, 79, 86–89; of methodological rules, 104; and observation and target theories, 47; and theory-ladenness, 38–43
Reichenbach, Hans, 97, 100n, 151
Relativism: definition of, viii; and evidential equivalence of theories, 54–56, 84; and holism, 111; and infallibilism, 86; and normative epistemology, 130, 134; scientific progress viewed by, 2; and social determination of belief, 146, 147, 161–63, 169; and theory-ladenness, 44
Relativity theory, 9–11, 113, 124, 125, 129, 130
Rhetoric, 163, 167, 168
Robustness of a theory, 60, 66
Rorty, Richard, 54, 86, 101, 112, 115, 121, 133, 136, 167
Rules, methodological. *See* Methodological rules

Scientific method: and convention, 88; and incommensurability, 135–38; and social determinants of belief, 153, 156, 161, 162; standards of success of, 94–107, 114, 116, 117; and theory choice, 34; and underdetermination, 66
Scientific progress. *See* Progress of science
Self-explanation, 52
Self-referentiality, 75, 88, 96, 169
Sellars, Wilfrid, 47, 59
Semantic equivalence of theories, 65
Shapin, Steven, 158

Skepticism, 14, 38, 86, 146
Social interests, 57, 146–60
Sociology of science, 135, 155
Standards: for evaluation of beliefs, 56, 93–119, 137; and incommensurability, 121; of a paradigm, 73, 74, 77
Straight rule of induction, 67
Strong relativism, 55, 56
Subjectivity, 98–100
Surprising predictions, 19, 25, 59, 60, 141–45

Target theory, 47, 48
Tessner, Peter, 158n
Testability, 5, 20
Tests of a theory: comparison of theories by, 14, 20, 21, 24–28, 31, 33, 140, 142–45; and confirming and positive instances of a theory, 62, 64; and holism, 57, 70; methodological guidelines for, 112, 113; and theory-ladenness, 38–40; and underdetermination, 48, 66. *See also* Well-tested theory
Theoretical entities, 8n, 10, 136–38
Theory (hypothesis): compatibility with observation of, 48, 50–55, 80; distinguished from observation, 37, 170; empirical success of, 166, 168; and holism, 70–72; incommensurability of, 121–24, 131, 135–37, 142–45; nondeducibility from observation of, 49, 56–58; and scientific progress, 3–34; social influences on, 147, 148, 151–57, 161; standards for acceptance of, 59–68, 93, 97, 100–107, 111, 112. *See also* Refutation of a theory; Tests of a theory
Theory-ladenness, 34–41, 44, 47, 48, 124
Translation, 123–33, 139–42
Truth: and incommensurability, 129; and nondeducibility of theories from observation, 56–58; and realism's account of science, 166; of relativists' assertions, 163; of theories, 49, 52, 53, 103, 105, 106, 143, 144; and underdetermination, 65, 67

Underdetermination, 48–54, 64–67; and equivalence of theories, 69; and incommensurability, 128, 131, 132, 135; and inductive rules, 59; and social determination of belief, 156, 160; and theory-ladenness, 34

Valuation, 104, 107, 112
Version of a paradigm, 72
Vortex theory of planetary motion, 15, 16

Weak relativism, 55, 56
Well-tested theory: and comparison of theories, 19, 28–30, 33, 143; and holism, 71, 79, 80; and underdetermination, 34, 54, 64, 65
Winch, Peter, 95n, 112, 113, 115, 118n, 119n
Wittgenstein, Ludwig, 112, 115, 116
Worldview, 13, 33, 71, 115, 121